Proceedings of the

Barcelona Postgrad Encounters on Fundamental Physics

October 17 – 19, 2012

Universitat de Barcelona, Spain

Edited by Daniel Fernández, Markus B. Fröb, Ivan Latella and Aldo Dector

Proceedings of the Barcelona Postgrad Encounters on Fundamental Physics

Editors: Daniel Fernández[1,a,b], Markus B. Fröb[2,a,b], Ivan Latella[3,a] and Aldo Dector[4,c]

[a]Departament de Física Fonamental

[b]Institut de Ciències del Cosmos (ICC)

[c]Departament d'Estructura i Constituents de la Matèria

 Universitat de Barcelona, C/ Martí i Franquès 1, 08028 Barcelona, Spain

[1]daniel@ffn.ub.edu, [2]mfroeb@ffn.ub.edu, [3]ilatella@ffn.ub.edu, [4]dector@ecm.ub.edu

We gratefully acknowledge financial support from the Facultat de Física, Universitat de Barcelona.

ISBN: 978-84-616-4293-9

Preface

During the last decades, the diverse areas of theoretical physics have evolved to become an interweaving whole. Even though they may still have different purposes and each one strives to acquire knowledge of the world at very different scales, they can no longer be understood as isolated fields. At the same time, international collaborations are standard practice and young investigators routinely go abroad for some time to do research together with renown physicists at different universities and research centers. On the other hand, the difficulty of understanding the laws of the Universe keeps increasing with every decade, making it vital to spread our cooperations as much as possible.

Thus, we must be aware of the importance to build strong networks at the international level. And the most efficient way to make these networks stronger is to start assembling ourselves as early as possible in our careers. The bonds created at this stage are the ones that will last longer throughout the years to come. While most workshops and conferences are conceived and organized by senior scientists, PhD and post-graduate students usually are involved in the details of the organizational work. Usually they also get a chance to present their work in afternoon sessions, after the main speakers which normally are also senior scientists have been given their share. In this way, as novices in the field of research those young investigators can learn about techniques and methods with their advantages and shortcomings, and benefit from the enormous wealth of knowledge that senior scientists usually have accumulated over the years. While the chance to meet and interact with experienced persons is of course priceless, we believe that a meeting only for young researchers also has its merits. The benefits of intense discussion between people who are in the same stages of their careers and facing the same problems are not to be underestimated.

In this sense, in our role of young researchers at the University of Barcelona, we decided to organize the *Barcelona Postgrad Encounters on Fundamental Physics*, as modest as it has been, to set an example. A precedent, maybe, for other institutions to follow in these footsteps. The support of our sponsor, the Faculty of Physics of the University of Barcelona, and in particular its dean, Joan Angel Padró, was essential for our idea to become reality. A denial of funding would have been understandable, especially in these times of economic crisis. After all, there exist almost no meetings like the one we were planning to organize in our faculty, and, without the motivation that was presented above, a workshop in which all the speakers are post-graduate students does not appear to be something appealing to allocate funds upon. But they listened and gave us the opportunity, and we are very thankful for it.

We would also like to thank the participants who answered our invitation. The success of this meeting is all due to their favorable response. Together they made the Barcelona Postgrad Encounters a memorable event for us, and we hope this was also the impression of everyone else. The organizing committee was very pleased to present an excellent panel of speakers to dissert on topics that gather enormous interest and activity. In these proceedings, we decided to split the received contributions into four sections: Gravitation, Condensed Matter, Particle and Radiation Physics, Supersymmetry and Supergravity as well as String theory and Holography. In the case where more than one author contributed to an entry, the speaker is underlined. As was mentioned previously, the different fields in theoretical physics are not disconnected at the fundamental level, and the reason for such divisions here is only to organize the book. The meeting displayed a high scientific level thanks to the excellent quality of the speakers, and to the adamant enthusiasm of the rest of the assistants. Drawing a conclusion, we believe this is proof of the good level of research that is being done nowadays by young physicists.

The programme spanned for three days, starting on October 17th, 2012. This opening day was entirely devoted to social interaction among the participants in a relaxed atmosphere. Talks were delivered on the 18th and 19th, following a scheme of 15-minute long participations. This time had to be limited due to the great number of participants, which exceeded our expectations. Since scientific cooperation was the main purpose of the meeting, there were also recesses in which the assistants could have open discussions.

As editors of these proceedings and members of the organizing committee, we want to use the opportunity to warmly thank everyone who made this conference possible. We hope that this meeting will be the first of many; if not in Barcelona, elsewhere in the world, and that PhD students and young post-doctorands can come together to forge collaborations and exchange their ideas. Wherever these future meetings take place, we wish them to have support from their institution as good as we had from ours.

Daniel Fernández

Markus B. Fröb

Ivan Latella

Aldo Dector

Contents

I

Gravitation

Nonlocal theories of gravity: the flat space propagator

Tirthabir Biswas[1,a], **Tomi Koivisto**[2,b] **and Anupam Mazumdar**[3,c]

[a]Physics Department, Loyola University, Campus Box 92, New Orleans, LA 70118

[b]Institutt for teoretisk astrofysikk, Universitetet i Oslo, 0315 Oslo, Norway

[c]Consortium for Fundamental Physics, Lancaster University, Lancaster, LA1 4YB, UK

E-mail: [1]tirthabir@gmail.com, [2]tomik@astro.uio.no, [3]a.mazumdar@lancaster.ac.uk

Abstract

It was recently found that there are classes of nonlocal gravity theories that are free of ghosts and singularities in their Newtonian limit [PRL **108** (2012), 031101]. In these proceedings, a detailed and pedagogical derivation of a main result, the flat space propagator for an arbitrary covariant metric theory of gravitation, is presented. The result is applied to analyse f(R) models, Gauß-Bonnet theory, Weyl-squared gravity and the potentially asymptotically free nonlocal theories.

1. Introduction

General Relativity (GR) predicts singularities and doesn't straightforwardly yield to quantisation. On the other hand, attempts to modify the theory are restricted by, besides phenomenological viability, theoretical consistency. By straying away from the Einstein-Hilbert action for the metric of space-time, one easily invites ghosts into the theory. While some specific higher derivative theories may be free of ghosts, they are not renormalisable, and vice versa [1].

However, nonlocal theories, featuring an infinite number of derivatives, might provide a way around this [2]. The ultraviolet singularities may then be smoothened out without introducing ghosts and while recovering GR predictions at small curvatures. Indeed, the promising attempts at quantum gravity such as string theory and loop quantum gravity exhibit nonlocality at some fundamental level. Phenomenologically nonlocal gravity has been recently applied to model pre-big bang cosmology [3, 4], inflation [5, 6], screening mechanisms [7, 8], dark energy [9, 10], structure formation [11, 12] and dark matter [13, 14]. Theoretical studies have considered renormalisability [15, 16] and black holes [17, 18].

In these proceedings, we will derive the flat space propagator for the most general metric theory of gravity, presented in [2]. In section 2 we write down the action and reduce it to a tractable form in the relevant Minkowski limit. Section 3 then introduces the formalism and the method to obtain the propagator. The result is applied in Section 4 for an analysis of several special cases of interest.

2. The most General Quadratic Action

To understand both the asymptotic behavior and the ghost free condition, what is relevant is the quadratic action of gravity. In other words if we look at fluctuations around the Minkowski background

$$g_{\mu\nu} = \eta_{\mu\nu} + h_{\mu\nu} \,, \tag{1}$$

then all we need to worry about are the terms that are quadratic in $h_{\mu\nu}$ in the action. Now, since in the Minkowski background $R_{\mu\nu\lambda\sigma}$ vanishes, everytime there is a Riemann tensor in the action it contributes an $\mathcal{O}(h)$ term in the action. Thereby we only need to analyse terms in the action that are products of at most two curvature terms, and the most general form for the action is given by

$$S_q = \int d^4x \sqrt{-g} R_{\mu_1\nu_1\lambda_1\sigma_1} \mathcal{D}^{\mu_1\nu_1\lambda_1\sigma_1}_{\mu_2\nu_2\lambda_2\sigma_2} R^{\mu_2\nu_2\lambda_2\sigma_2} \,, \tag{2}$$

where $\mathcal{D}$ is a differential operator containing covariant derivatives and $\eta_{\mu\nu}$. We note that if there is a differential operator acting on the left Riemann tensor as well, one can always recast that into the above form using integration by parts.

Since the operator $\mathcal{D}$ can only have covariant derivatives and the Minkowski metric, one can actually write down the most general action S_q explicitly:

$$\begin{aligned}
S_q = \int d^4x \sqrt{-g} \Big[& RF_1(\Box)R + RF_2(\Box)\nabla_\mu\nabla_\mu R^{\mu\nu} + R_{\mu\nu}F_3(\Box)R^{\mu\nu} + R^\nu_\mu F_4(\Box)\nabla_\nu\nabla_\lambda R^{\mu\lambda} \\
& + R^{\lambda\sigma}F_5(\Box)\nabla_\mu\nabla_\sigma\nabla_\nu\nabla_\lambda R^{\mu\nu} + RF_6(\Box)\nabla_\mu\nabla_\nu\nabla_\lambda\nabla_\sigma R^{\mu\nu\lambda\sigma} + R_{\mu\lambda}F_7(\Box)\nabla_\nu\nabla_\sigma R^{\mu\nu\lambda\sigma} \\
& + R^\rho_\lambda F_8(\Box)\nabla_\mu\nabla_\sigma\nabla_\nu\nabla_\rho R^{\mu\nu\lambda\sigma} + R^{\mu_1\nu_1}F_9(\Box)\nabla_{\mu_1}\nabla_{\nu_1}\nabla_\mu\nabla_\nu\nabla_\lambda\nabla_\sigma R^{\mu\nu\lambda\sigma} \\
& + R_{\mu\nu\lambda\sigma}F_{10}(\Box)R^{\mu\nu\lambda\sigma} + R^\rho_{\mu\nu\lambda}F_{11}(\Box)\nabla_\rho\nabla_\sigma R^{\mu\nu\lambda\sigma} \\
& + R_{\mu\rho_1\nu\sigma_1}F_{12}(\Box)\nabla^{\rho_1}\nabla^{\sigma_1}\nabla_\rho\nabla_\sigma R^{\mu\rho\nu\sigma} + R^{\nu_1\rho_1\sigma_1}_\mu F_{13}(\Box)\nabla_{\rho_1}\nabla_{\sigma_1}\nabla_{\nu_1}\nabla_\nu\nabla_\rho\nabla_\sigma R^{\mu\nu\lambda\sigma} \\
& + R^{\mu_1\nu_1\rho_1\sigma_1}F_{14}(\Box)\nabla_{\rho_1}\nabla_{\sigma_1}\nabla_{\nu_1}\nabla_{\mu_1}\nabla_\mu\nabla_\nu\nabla_\rho\nabla_\sigma R^{\mu\nu\lambda\sigma} \Big] \,.
\end{aligned} \tag{3}$$

Using the antisymmetric properties of the Riemann tensor,

$$R_{(\mu\nu)\rho\sigma} = R_{\mu\nu(\rho\sigma)} = 0, \tag{4}$$

and the Jacobi identity

$$\nabla_\alpha R_{\mu\nu\beta\gamma} + \nabla_\gamma R_{\mu\nu\alpha\beta} + \nabla_\beta R_{\mu\nu\gamma\alpha} = 0, \tag{5}$$

one finds after patient index manipulation that the above action reduces to

$$
\begin{aligned}
S_q = \int \mathrm{d}^4 x \sqrt{-g} \Big[& R F_1(\Box) R + R_{\mu\nu} F_3(\Box) R^{\mu\nu} + R F_6(\Box) \nabla_\mu \nabla_\nu \nabla_\lambda \nabla_\sigma R^{\mu\nu\lambda\sigma} \\
& + R_{\mu\nu\lambda\sigma} F_{10}(\Box) R^{\mu\nu\lambda\sigma} + R_\mu^{\nu_1\rho_1\sigma_1} F_{13}(\Box) \nabla_{\rho_1} \nabla_{\sigma_1} \nabla_{\nu_1} \nabla_\nu \nabla_\rho \nabla_\sigma R^{\mu\nu\lambda\sigma} \\
& + R^{\mu_1\nu_1\rho_1\sigma_1} F_{14}(\Box) \nabla_{\rho_1} \nabla_{\sigma_1} \nabla_{\nu_1} \nabla_{\mu_1} \nabla_\mu \nabla_\nu \nabla_\rho \nabla_\sigma R^{\mu\nu\lambda\sigma} \Big].
\end{aligned}
\tag{6}
$$

So we got rid of 8 of the 14 terms already in a curved background.

2.1. Linearised Action

Our next task is to obtain the quadratic (in $h_{\mu\nu}$) free part of the above action. A very important simplification occurs when we realise that the two h-dependent terms must come from the two curvature terms present. In other words the covariant derivatives must take on the Minkowski values, and we can commute them freely. We then observe that the F_6, F_{13} and F_{14} terms in the action (6) become irrelevant in this limit due to the symmetry of the derivative operations contracting the antisymmetric index pairs of the Riemann tensor (4). The linearised action contains in the end only

$$S_q = \int \mathrm{d}^4 x \left[R F_1(\Box) R + R_{\mu\nu} F_3(\Box) R^{\mu\nu} + R_{\mu\nu\lambda\sigma} F_{10}(\Box) R^{\mu\nu\lambda\sigma} \right]. \tag{7}$$

Furthermore, below it will become clear that there are essentially only two free functions that determine the properties of the theory in this limit.

Now our next task is to substitute the linearised expressions of the curvatures in terms of $h_{\mu\nu}$:

$$
R_{\mu\nu\lambda\sigma} = \frac{1}{2}(\partial_{[\lambda}\partial_{\nu}h_{\mu\sigma]} - \partial_{[\lambda}\partial_{\mu}h_{\nu\sigma]}),
$$

$$
R_{\mu\nu} = \frac{1}{2}(\partial_\sigma\partial_{(\nu}h^\sigma_{\mu)} - \partial_\nu\partial_\mu h - \Box h_{\mu\nu}), \quad R = \partial_\nu\partial_\mu h^{\mu\nu} - \Box h.
$$

As is obvious, many of the terms simplify and combine. By considering all possible consistent contractions, one can deduce that all the terms eventually have to produce an action of the following form:

$$S_q = -\int d^4x \left[\frac{1}{2} h_{\mu\nu} \Box a(\Box) h^{\mu\nu} + h_\mu^\sigma b(\Box) \partial_\sigma \partial_\nu h^{\mu\nu} + hc(\Box) \partial_\mu \partial_\nu h^{\mu\nu} \right.$$
$$\left. + \frac{1}{2} h \Box d(\Box) h + h^{\lambda\sigma} \frac{f(\Box)}{\Box} \partial_\sigma \partial_\lambda \partial_\mu \partial_\nu h^{\mu\nu} \right], \tag{8}$$

where we have defined the functions $a(\Box)$, $b(\Box)$, $c(\Box)$ and $d(\Box)$ in such a way that they reduce in the appropriate limit to the constants a, b, c and d used by van Nieuwenhuizen [19]. The function $f(\Box)$ appears only in higher order theories.

We will now compute all of the terms in the original action (7) individually. The first piece gives

$$RF_1(\Box)R = hF_1\Box^2 h + h^{\lambda\sigma} F_1 \partial_\sigma \partial_\lambda \partial_\mu \partial_\nu h^{\mu\nu} - hF_1\Box \partial_\mu \partial_\nu h^{\mu\nu} - h^{\mu\nu} F_1 \Box \partial_\mu \partial_\nu h.$$

The third and fourth terms in this case can be combined as follows. Ignoring surface terms it is always possible to commute through the local $F(\Box)$ terms[1], and we get

$$RF_1(\Box)R = F_1(\Box) \left[h\Box^2 h + h^{\lambda\sigma} \partial_\sigma \partial_\lambda \partial_\mu \partial_\nu h^{\mu\nu} - 2h\Box \partial_\mu \partial_\nu h^{\mu\nu} \right].$$

For the two other relevant terms in the (7) we obtain in the similar way:

$$R_{\mu\nu} F_3(\Box) R^{\mu\nu} = F_3(\Box) \left[\frac{1}{4} h\Box^2 h + \frac{1}{4} h_{\mu\nu} \Box^2 h^{\mu\nu} - \frac{1}{2} h_\mu^\sigma \Box \partial_\sigma \partial_\nu h^{\mu\nu} \right.$$
$$\left. - \frac{1}{2} h\Box \partial_\mu \partial_\nu h^{\mu\nu} + \frac{1}{2} h^{\lambda\sigma} \partial_\sigma \partial_\lambda \partial_\mu \partial_\nu h^{\mu\nu} \right];$$
$$R_{\mu\nu\lambda\sigma} F_{10}(\Box) R^{\mu\nu\lambda\sigma} = F_{10}(\Box) \left[h_{\mu\nu} \Box^2 h^{\mu\nu} + h^{\lambda\sigma} \partial_\sigma \partial_\lambda \partial_\mu \partial_\nu h^{\mu\nu} - 2h_\mu^\sigma \Box \partial_\sigma \partial_\nu h^{\mu\nu} \right].$$

It remains to relate these terms to the five combinations appearing in action (8).

2.2. The coefficients in terms of $F_i(\Box)$

In the above section we have calculated the contribution of the higher derivative modifications (7) to the action (8). We also need to include the contribution from the Einstein-Hilbert term, so that the full action we consider is

$$S = \int d^4x \sqrt{-g} R + S_q. \tag{9}$$

---

[1]For non-polynomial terms that is not clear.

We then eventually obtain

$$a(\Box) = 1 - \frac{1}{2}F_3(\Box)\Box - 2F_{10}(\Box)\Box,$$

$$b(\Box) = -1 + \frac{1}{2}F_3(\Box)\Box + 2F_{10}(\Box)\Box,$$

$$c(\Box) = 1 + 2F_1(\Box)\Box + \frac{1}{2}F_3(\Box)\Box, \tag{10}$$

$$d(\Box) = -1 - 2F_1(\Box)\Box - \frac{1}{2}F_3(\Box)\Box,$$

$$f(\Box) = -2F_1(\Box)\Box - F_3(\Box)\Box - 2F_{10}(\Box)\Box.$$

From the above expressions we observe the following interesting relations[2]

$$a + b = 0, \quad c + d = 0, \quad b + c + f = 0, \tag{11}$$

so that we are really left with two independent arbitrary functions. This can be understood as a consequence of the Bianchi identities, as will be shortly clarified.

2.3. Field Equations and Bianchi identities

The field equations can be derived straightforwardly by varying the action (8):

$$a(\Box)\Box h_{\mu\nu} + b(\Box)\partial_\sigma\partial_{(\nu}h^\sigma_{\mu)} + c(\Box)(\eta_{\mu\nu}\partial_\rho\partial_\sigma h^{\rho\sigma} + \partial_\mu\partial_\nu h)$$
$$+ \eta_{\mu\nu}d(\Box)\Box h + f(\Box)\Box^{-1}\partial_\sigma\partial_\lambda\partial_\mu\partial_\nu h^{\lambda\sigma} = -\kappa\tau_{\mu\nu}. \tag{12}$$

The matter side is conserved by the stress energy conservation and the geometric part because of the generalised Bianchi identities [20] due to diffeomorphism invariance. Thus

$$-\kappa\tau\nabla_\mu\tau^\mu_\nu = 0 = (c + d)\Box\partial_\nu h + (a + b)\Box h^\mu_{\nu,\mu} + (b + c + f)h^{\alpha\beta}_{,\alpha\beta\nu}. \tag{13}$$

It is then clear why (11) had to hold.

The above field equations can be written in the form

$$\Pi^{-1\lambda\sigma}_{\mu\nu}h_{\lambda\sigma} = \kappa\tau_{\mu\nu}, \tag{14}$$

where $\Pi^{-1\lambda\sigma}_{\mu\nu}$ is the inverse propagator. To compute the propagator, we need to learn to deal with spin projector operators.

---

[2]An immediate observation we can make is that one cannot construct the Fierz-Pauli term, for which $a = -d \sim m^2$ and $b = c = 0$ from an action like (3). Massive gravity is not among the metric theories we consider here, but inherently bimetric.

3. Propagators

In this section we are going to derive the propagators for the field equations (12). The basic algorithm is as follows [19]: First, we express the field equations in the form (14), where the inverse propagator Π^{-1} is expressed in terms of six operators $\mathcal{P}_i$ (to be specified shortly):

$$\Pi^{-1} = \sum_{i=1}^{6} C_i \mathcal{P}_i \qquad (15)$$

In the momentum space description the coefficients C_i are scalars which can only depend on k^2. Finding the suitable operators, it is possible to decompose the field equations into a decoupled set of equations of motion for the relevant degrees of freedom. These are then conveniently invertible.

3.1. Spin projector operators

Let us introduce

$$\begin{aligned}
\mathcal{P}^2 &= \frac{1}{2}(\theta_{\mu\rho}\theta_{\nu\sigma} + \theta_{\mu\sigma}\theta_{\nu\rho}) - \frac{1}{3}\theta_{\mu\nu}\theta_{\rho\sigma}, \\
\mathcal{P}^1 &= \frac{1}{2}(\theta_{\mu\rho}\omega_{\nu\sigma} + \theta_{\mu\sigma}\omega_{\nu\rho} + \theta_{\nu\rho}\omega_{\mu\sigma} + \theta_{\nu\sigma}\omega_{\mu\rho}), \\
\mathcal{P}^0_s &= \frac{1}{3}\theta_{\mu\nu}\theta_{\rho\sigma}, \quad \mathcal{P}^0_w = \omega_{\mu\nu}\omega_{\rho\sigma}, \\
\mathcal{P}^0_{sw} &= \frac{1}{\sqrt{3}}\theta_{\mu\nu}\omega_{\rho\sigma}, \quad \mathcal{P}^0_{ws} = \frac{1}{\sqrt{3}}\omega_{\mu\nu}\theta_{\rho\sigma},
\end{aligned} \qquad (16)$$

where the transversal and longitudinal projectors in the momentum space are respectively

$$\theta_{\mu\nu} = \eta_{\mu\nu} - \frac{k_\mu k_\nu}{k^2}, \qquad \omega_{\mu\nu} = \frac{k_\mu k_\nu}{k^2}.$$

Note that the operators $\mathcal{P}^i$ are in fact 4-rank tensors, $\mathcal{P}^i_{\mu\nu\rho\sigma}$, but we have suppressed the index notation here.

Out of the six operators four of them, $\{\mathcal{P}^2, \mathcal{P}^1, \mathcal{P}^0_s, \mathcal{P}^0_w\}$, form a complete set of projection operators:

$$\mathcal{P}^i_a \mathcal{P}^j_b = \delta^{ij}\delta_{ab}\mathcal{P}^i_a \quad \text{and} \quad \mathcal{P}^2 + \mathcal{P}^1 + \mathcal{P}^0_s + \mathcal{P}^0_w = 1, \qquad (17)$$

as one can easily verify. These projection operators together represent the six field degrees of freedom, the additional four fields in a symmetric tensor field, as usual, being gauge degrees of freedom. $\mathcal{P}^2$ and $\mathcal{P}^1$ represent transverse and traceless spin-2 and spin-1 degrees, accounting for four field degrees of freedom, while $\mathcal{P}^0_s$, $\mathcal{P}^0_w$

represent the spin-0 scalar multiplets. In addition to the above four spin operators we have $\mathcal{P}^0_{sw}$ and $\mathcal{P}^0_{ws}$ which can potentially mix the two scalar multiplets. In particular, we have that

$$\mathcal{P}^0_{ij}\mathcal{P}^0_k = \delta_{jk}\mathcal{P}^0_{ij}, \quad \mathcal{P}^0_{ij}\mathcal{P}^0_{kl} = \delta_{il}\delta_{jk}\mathcal{P}^0_k, \quad \mathcal{P}^0_k\mathcal{P}^0_{ij} = \delta_{ik}\mathcal{P}^0_{ij}, \tag{18}$$

as one may again easily check.

From (15) and (17) it trivially follows that we can write (14) as

$$\sum_{i=1}^{6} C_i \mathcal{P}_i h = \kappa(\mathcal{P}^2 + \mathcal{P}^1 + \mathcal{P}^0_s + \mathcal{P}^0_w)\tau. \tag{19}$$

By multiplying with the different projector operators on either side of the equation we can now obtain the decoupled field equations for the different spin multiplets.

3.2. Inverting the field equations

Having outlined the algorithm for finding the propagators, let us now proceed to obtain them in our model specified by the action (8). We need to express all the operators in (12) in terms of the operators $\mathcal{P}^i$. Let us start with

$$\begin{aligned}
\eta_{\mu\nu}d(\Box)h \to d(-k^2)\eta_{\mu\nu}\eta^{\rho\sigma}h_{\rho\sigma} &= d(-k^2)(\theta_{\mu\nu} + \omega_{\mu\nu})(\theta^{\rho\sigma} + \omega^{\rho\sigma})h_{\rho\sigma} \\
&= d(-k^2)(\theta_{\mu\nu}\theta^{\rho\sigma} + \omega_{\mu\nu}\omega^{\rho\sigma} + \theta_{\mu\nu}\omega^{\rho\sigma} + \omega_{\mu\nu}\omega^{\rho\sigma})h_{\rho\sigma} \\
&= d(-k^2)[3\mathcal{P}^0_s + \mathcal{P}^0_w + \sqrt{3}(\mathcal{P}^0_{sw} + \mathcal{P}^0_{ws})]h.
\end{aligned}$$

One can continue in an analogous fashion to obtain the projector decomposition of all the operators appearing in the field equations (12). For the first three terms we then obtain

$$a(\Box)h_{\mu\nu} \to a(-k^2)\left[\mathcal{P}^2 + \mathcal{P}^1 + \mathcal{P}^0_s + \mathcal{P}^0_w\right]h,$$

$$b(\Box)\partial_\sigma\partial_{(\nu}h^\sigma_{\mu)} \to -b(-k^2)k^2\left[\mathcal{P}^1 + 2\mathcal{P}^0_w\right]h,$$

$$c(\Box)(\eta_{\mu\nu}\partial_\rho\partial_\sigma h^{\rho\sigma} + \partial_\mu\partial_\nu h) \to -c(-k^2)k^2\left[2\mathcal{P}^0_w + \sqrt{3}\left(\mathcal{P}^0_{sw} + \mathcal{P}^0_{ws}\right)\right]h.$$

While all the above operators appear in two derivative generalisations of gravity and were discussed in [19], the f term in (12) is specific to higher derivative theories. Its decomposition is rather simple

$$f(\Box)\partial^\sigma\partial^\rho\partial_\mu\partial_\nu h_{\rho\sigma} \to f(-k^2)k^\sigma k^\rho k_\mu k_\nu h_{\rho\sigma} = f(-k^2)k^4\omega_{\mu\nu}\omega^{\rho\sigma} = f(-k^2)k^4\mathcal{P}^0_w.$$

We are now ready to write down the projected field equations, and the corresponding propagators. By acting with $\mathcal{P}^2$ on (19) we find

$$ak^2\mathcal{P}^2h = \kappa\mathcal{P}^2\tau \Rightarrow \mathcal{P}^2h = \kappa\left(\frac{\mathcal{P}^2}{ak^2}\right)\tau. \tag{20}$$

Similarly, acting with $\mathcal{P}^1$, one finds

$$(a+b)k^2\mathcal{P}^1 h = \kappa \mathcal{P}^1 \tau. \tag{21}$$

Rather interestingly, since recalling Eq. (11) we know that $a+b=0$, this implies that there are in fact no vector degrees of freedom, and accordingly the stress-energy tensor must have no vectorial part: $\mathcal{P}^1 \tau = 0$.

Next let us look at the scalar multiplets. By acting $\mathcal{P}^0_s$ and $\mathcal{P}^0_w$ on (19) we obtain

$$(a+3d)k^2\mathcal{P}^0_s h + (c+d)k^2\sqrt{3}\mathcal{P}^0_{sw}h = \kappa\mathcal{P}^0_s\tau, \tag{22}$$

$$(c+d)k^2\sqrt{3}\mathcal{P}^0_{ws}h + (a+2b+2c+d+f)k^2\mathcal{P}^0_w h = \kappa\mathcal{P}^0_w\tau. \tag{23}$$

As we see, in principle, the scalar multiplets are coupled. However, by applying the projector $\mathcal{P}^0_w$ on equation (22) or the projector $\mathcal{P}^0_s$ on equation (23) from the right hand side, one sees that $c+d=0$, in accordance with (11). The scalars decouple and one can now straightforwardly invert the field equations to obtain the propagators:

$$(a+3d)k^2\mathcal{P}^0_s h = \kappa\mathcal{P}^0_s\tau \Rightarrow \mathcal{P}^0_s h = \kappa\frac{\mathcal{P}^0_s}{(a+3d)k^2}\tau \text{ and} \tag{24}$$

$$(a+2b+2c+d+4f)k^2\mathcal{P}^0_w h = \kappa\mathcal{P}^0_w\tau \Rightarrow \mathcal{P}^0_w h = \kappa\frac{\mathcal{P}^0_w}{(a+2b+2c+d+f)k^2}\tau, \tag{25}$$

respectively. The denominator corresponding to the $\mathcal{P}^0_w$ projector vanishes. So there is no w-multiplet, but the s-multiplet picks up a nontrivial propagator.

To finally summarise:

$$\Pi = \frac{\mathcal{P}^2}{ak^2} + \frac{\mathcal{P}^0_s}{(a-3c)k^2}. \tag{26}$$

We have thus arrived at the main result of [2].

4. Applications to special cases

In this section, we consider the implications of the result (26) to some special cases.

4.1. General Relativity

Since we want to recover GR in the infrared, we require from any viable theory that

$$a(0) = c(0) = -b(0) = -d(0) = 1, \tag{27}$$

corresponding to the GR values. In GR these functions are the same constants for any Fourier mode. The above condition ensures that as $k^2 \to 0$, we have only the physical graviton propagator,

$$\lim_{k^2 \to 0} \Pi = (\mathcal{P}^2/k^2) - (\mathcal{P}^0_s/2k^2) \equiv \Pi_{\text{GR}}. \tag{28}$$

There is a crucial subtlety one should observe here. Although the $\mathcal{P}^0_s$ residue at $k^2 = 0$ is negative, that is a benign ghost. In fact, $\mathcal{P}^0_s$ has precisely the right coefficient to cancel the unphysical longitudinal degrees of freedom in the spin-2 part [19].

4.2. Gauß-Bonnet gravity

Let us consider the theory $\mathcal{L} = R + \alpha(\Box)G$, where G is the Gauß-Bonnet invariant $G = R^2 - 4R_{\mu\nu}R^{\mu\nu} + R_{\mu\nu\rho\sigma}R^{\mu\nu\rho\sigma}$ and the function $\alpha(\Box)$ in the simplest case can be just a constant coefficient. As is well known, in four dimensions the Gauß-Bonnet term is a topological invariant that does not contribute to the gravitational field equations. Therefore it is not a surprise that it doesn't introduce any modifications to the propagator either. In Eq. (3) we have now $F_1 = \alpha$, $F_3 = -4\alpha$ and $F_{10} = \alpha$. Regardless of α we then see from Eqs. (10) that $a = c = -b = -d = 1$, and thus the properties of the theory are identical to GR.

4.3. $\mathcal{L}(R)$ gravity

The $\mathcal{L}(R)$ gravity is a popular subject of study. For our purposes here, it is enough to consider the expansion of the lagrangian around flat space,

$$\mathcal{L}(R) = \mathcal{L}(0) + \mathcal{L}'(0)R + \frac{1}{2}\mathcal{L}''(0)R^2 + \cdots. \tag{29}$$

The first term one identifies with the cosmological constant, $\mathcal{L}(0) = -2\kappa^{-1}\Lambda$, and the second term should reduce to the Einstein-Hilbert term in a viable theory, $\mathcal{L}'(0) = 1$. The relevant modification of the theory is then given by the quadratic part. Since only F_1 is now nonzero in (3), we readily see from (10) that then $a = -b = 1$, $c = -d = 1 - \mathcal{L}''(0)\Box$ and $f = -\mathcal{L}''(0)\Box$. The propagator is thus

$$\Pi = \frac{\mathcal{P}^2}{k^2} - \frac{\mathcal{P}^0_s}{2k^2(1 + 3\mathcal{L}''(0)k^2)}. \tag{30}$$

The scalar part of the propagator is modified. Since these theories are a specific class of scalar-tensor theories, we expect an extra scalar degree of freedom. Its appearance can be made transparent by rewriting the above result as

$$\Pi = \Pi_{\text{GR}} + \frac{1}{2}\frac{\mathcal{P}^0_s}{k^2 + m^2}, \quad m^2 = \frac{1}{3\mathcal{L}''(0)}. \tag{31}$$

21

Indeed, the $\mathcal{L}(R)$ correction entails an additional spin-0 particle which is nontachyonic as long as[3] $\mathcal{L}''(0) > 0$. One also sees that though these theories are classically viable, they cannot improve the ultraviolet properties of GR since the graviton propagator retains its form.

4.4. Conformally invariant gravity

As an example of a ghastly theory, let us consider the Weyl squared gravity. The Weyl tensor is defined as

$$C_{\mu\nu\rho\sigma} = R_{\mu\nu\rho\sigma} + \frac{R}{6}\left(g_{\mu\rho}g_{\nu\sigma} - g_{\mu\sigma}g_{\nu\rho}\right) - \frac{1}{2}\left(g_{\mu\rho}R_{\nu\sigma} - g_{\mu\sigma}R_{\nu\rho} - g_{\nu\rho}R_{\mu\sigma} + g_{\nu\sigma}R_{\mu\rho}\right).$$

The theory is then specified by the conformally invariant Weyl-squared term, $\mathcal{L} = R - \frac{1}{m^2}C^2$, where m is the mass scale at which the correction becomes relevant. It is straightforward to compute that

$$C^2 = R_{\mu\nu\rho\sigma}R^{\mu\nu\rho\sigma} - 2R_{\mu\nu}R^{\mu\nu} + \frac{1}{3}R^2, \tag{32}$$

from which we quickly infer, using again Eqs.(3,10) that now $a = -b = 1 - (k/m)^2$, $c = -d = 1 - (k/m)^2/3$ and $f = -2(k/m)^2/3$. We obtain the propagator with a double pole for the graviton:

$$\Pi = \frac{\mathcal{P}^2}{k^2\left(1 - (k/m)^2\right)} - \frac{\mathcal{P}^0_s}{2k^2} = \Pi_{\text{GR}} - \frac{\mathcal{P}^2}{k^2 + m^2}. \tag{33}$$

From the latter form of the propagator it is obvious that the theory contains an extra spin-2 degree of freedom with respect to GR. Moreover, the extra contribution always comes with the wrong sign: this is the Weyl ghost[4].

4.5. Asymptotically free gravity

Finally, we show how the ultraviolet properties of GR are improved via a nonlocal extension. Just for simplicity, let us restrict to the special class of theories with $f = 0$. From (11) we then see that $a = c = -b = -d$. This means that we are essentially left with just a single free function

$$a(\Box) = 1 - \frac{1}{2}F_3(\Box)\Box - 2F_{10}(\Box)\Box. \tag{34}$$

---

[3]For an alternative derivation and generalisation of this stability condition, see [21].
[4]However perhaps the negative norm states can be consistently projected out of the Hilbert space [22].

We obtain a very simple expression for the propagator:

$$\Pi = \frac{1}{k^2 a(-k^2)} \left(\mathcal{P}^2 - \frac{1}{2}\mathcal{P}^0_s \right) = \frac{1}{a(-k^2)} \Pi_{\text{GR}} . \tag{35}$$

Thus, the GR propagator is now modulated by the k-dependent function $a(\Box)$. We now realise that as long as $a(\Box)$ has no zeroes, these theories contain no new states as compared to GR, and only modify the physical graviton propagator. Polynomial functions would correspond to higher - but finite - order gravity, and would inevitably result in new pathological states. This can be avoided in nonlocal, i.e. infinite order higher derivative theories. Furthermore, by choosing $a(k^2)$ to be a suitable entire function we can indeed tame the behavior of the ultraviolet gravitons. A simple example can be provided by $a = \exp[(k/M)^2]$, where M is a mass scale at which the nonlocal modifications become important. The integrals over the propagator quickly tend to zero at high momenta $k > M$ and we expect finite results from physical calculations (note though that light-like momenta do not receive damping).

5. Conclusions

Having derived the main result (26), we considered its implications in some special cases. We readily reproduced the known results: while GR and Gauß-Bonnet theory share the same field content, $f(R)$ gravity has an extra healthy scalar and Weyl gravity an extra pathological spin-2 field. New classes of nonlocal theories were found, that are both unitary and devoid of singularities. The ongoing further work includes the generalisation of the result to curved backgrounds.

Acknowledgements

TK would like to thank Erik Gerwick and Alex Koshelev for their contributions to these calculations and Danielle Wills, Sergey Vernov and Nicola Tamanini for insightful discussions. TB is supported by the Louisiana Board of Regents, TK by the Research Council of Norway and AM by the STFC grant ST/J000418/1.

References

[1]   K. Stelle, "Renormalization of Higher Derivative Quantum Gravity", *Phys. Rev.* **D16** (1977), 953.

[2]   T. Biswas et al., "Towards singularity and ghost free theories of gravity", *Phys. Rev. Lett.* **108** (2012), 031101.

[3]   T. Biswas, A. Mazumdar, W. Siegel, "Bouncing universes in string-inspired gravity", *JCAP* **0603** (2006), 009.

[4] T. Biswas, T. Koivisto, A. Mazumdar, "Towards a resolution of the cosmological singularity in non-local higher derivative theories of gravity", *JCAP* **1011** (2010), 008.

[5] S. Capozziello et al., "Accelerating cosmologies from non-local higher-derivative gravity", *Phys. Lett.* **B671** (2009), 193.

[6] T. Biswas et al., "Stable bounce and inflation in non-local higher derivative cosmology", *JCAP* **1208** (2012), 024.

[7] N. Arkani-Hamed et al., "Nonlocal modification of gravity and the cosmological constant problem" (2002), arXiv:hep-th/0209227.

[8] Y.-l. Zhang, M. Sasaki, "Screening of cosmological constant in non-local cosmology", *Int. J. Mod. Phys.* **D21** (2012), 1250006.

[9] S. Deser, R. Woodard, "Nonlocal Cosmology", *Phys. Rev. Lett.* **99** (2007), 111301.

[10] T. Koivisto, "Dynamics of Nonlocal Cosmology", *Phys. Rev.* **D77** (2008), 123513.

[11] T. S. Koivisto, "Newtonian limit of nonlocal cosmology", *Phys. Rev.* **D78** (2008), 123505.

[12] S. Park, S. Dodelson, "Structure formation in a nonlocally modified gravity model", *Phys. Rev.* **D87** (2013), 024003.

[13] M. Soussa, R. P. Woodard, "A Nonlocal metric formulation of MOND", *Class. Quant. Grav.* **20** (2003), 2737.

[14] H.-J. Blome et al., "Nonlocal Modification of Newtonian Gravity", *Phys. Rev.* **D81** (2010), 065020.

[15] J. Moffat, "Ultraviolet Complete Quantum Gravity", *Eur. Phys. J. Plus* **126** (2011), 43.

[16] L. Modesto, "Towards a finite quantum supergravity" (2012), arXiv:1206.2648.

[17] L. Modesto, J. W. Moffat, P. Nicolini, "Black holes in an ultraviolet complete quantum gravity", *Phys. Lett.* **B695** (2011), 397.

[18] P. Nicolini, "Nonlocal and generalized uncertainty principle black holes" (2012), arXiv:1202.2102.

[19] P. Van Nieuwenhuizen, "On ghost-free tensor lagrangians and linearized gravitation", *Nucl. Phys.* **B60** (1973), 478.

[20] T. Koivisto, "Covariant conservation of energy momentum in modified gravities", *Class. Quant. Grav.* **23** (2006), 4289.

[21] L. Amendola, K. Enqvist, T. Koivisto, "Unifying Einstein and Palatini gravities", *Phys. Rev.* **D83** (2011), 044016.

[22] P. D. Mannheim, "Making the Case for Conformal Gravity", *Found. Phys.* **42** (2012), 388.

Can one observe effects of quantum gravity in the cosmic microwave background?

Manuel Krämer

Institut für Theoretische Physik, Universität zu Köln,
Zülpicher Straße 77, 50937 Köln, Germany

E-mail: mk@thp.uni-koeln.de

Abstract
The correct theory of quantum gravity can only be found by looking for observational effects. Here, we use the framework of canonical quantum gravity to calculate quantum-gravitational contributions to the power spectrum of cosmological perturbations in an inflationary universe. The resulting modification is too weak to be observable in the anisotropies of the cosmic microwave background radiation, but we can deduce a constraint on the energy scale of inflation.

1. Introduction

For more than 70 years, physicists have been trying to find a quantum theory of gravity. While up to today various approaches of such a theory of quantum gravity have been developed, none of them has lead to a testable prediction and therefore reached the status of a well-established theory yet. The lack of testable predictions is a consequence of the general smallness of quantum-gravitational effects, which is due to the expectation that situations in which both gravity and quantum theory play a role must involve energies in the region of the Planck scale, which amounts to about 10^{19} GeV. This rules out any accelerator experiment to probe quantum gravity and in principle limits the search for observational effects of quantum gravity to black hole physics and very early universe cosmology.

Here, we shall focus to look for potentially observable quantum-gravitational effects in the anisotropy spectrum of the cosmic microwave background (CMB) radiation, which has turned out to provide us with a plethora of information about the physics of the very early universe, such that it is one of the best candidates to look for effects of quantum gravity.

The framework we have chosen is *Quantum Geometrodynamics*, which is a direct canonical quantization of general relativity. Even though it is unlikely that this approach to quantum gravity turns out to be the final answer, it can be used as an effective theory because in the semiclassical limit one recovers Einstein's equations from it.

The central equation of *Quantum Geometrodynamics* is the *Wheeler-DeWitt equation* and our aim here is to calculate the dominant quantum-gravitational contribution to the power spectrum of cosmological perturbations in an inflationary universe by performing a semiclassical approximation to the Wheeler-DeWitt equation of a suitable quantum-cosmological model. The resulting modification of the power spectrum can then be translated to the CMB anisotropy spectrum.

This conference contribution is based on the papers [1] and [2].

2. The quantum-cosmological model

The quantum-cosmological model we choose to derive the power spectrum of perturbations in an inflationary universe is very simple, we only consider perturbations of a scalar field ϕ, which plays the role of the inflaton. Even though we neglect the perturbations of the metric itself, the model is suitable to give a first estimate of the magnitude of quantum-gravitational effects.

We assume the background universe to be a flat Friedmann-Lemaître universe with scale factor $a \equiv \exp(\alpha)$. The inflaton potential is set to be $\mathcal{V}(\phi) = \frac{1}{2} m^2 \phi^2 \approx$ const. for definiteness. Furthermore, the slow-roll approximation in the form of $\dot{\phi}^2 \ll |\mathcal{V}(\phi)|$ is demanded to hold. In order to decompose the perturbations of the inflaton field into Fourier modes, we assume for simplicity that the space is compact and that the spectrum of the wave vector k, $k := |\mathbf{k}|$ – which we define to be dimensionless – is discrete. Therefore we can write $\delta\phi(\mathbf{x}, t) = \sum_k f_k(t) e^{i\mathbf{k}\cdot\mathbf{x}}$.

We set $\hbar = c = 1$ and redefine the Planck mass as $m_{\mathrm{P}} := \sqrt{3\pi/2G} \approx 2.65 \times 10^{19}$ GeV for later convenience. The quasi-static inflationary Hubble parameter is denoted as H. Based on [3], we can then write out the Wheeler-DeWitt equation for each of the modes f_k as follows

$$[\mathcal{H}_0 + \mathcal{H}_k]\Psi_k(\alpha, f_k) = 0, \tag{1}$$

where $\mathcal{H}_0$ is the Hamiltonian of the background and the $\mathcal{H}_k$ are the Hamiltonians of the perturbation modes:

$$\mathcal{H}_0 = \frac{e^{-3\alpha}}{2}\left[\frac{1}{m_{\mathrm{P}}^2}\frac{\partial^2}{\partial\alpha^2} + e^{6\alpha}m_{\mathrm{P}}^2 H^2\right], \quad \mathcal{H}_k = \frac{e^{-3\alpha}}{2}\left[-\frac{\partial^2}{\partial f_k^2} + \left(k^2 e^{4\alpha} + m^2 e^{6\alpha}\right)f_k^2\right]. \tag{2}$$

3. The semiclassical approximation and the derivation of the power spectrum

Our aim is to find quantum-gravitational correction terms to the expressions that are used to calculate the primordial power spectrum of cosmological perturbations. Hence, it is not necessary to solve the full Wheeler-DeWitt equation (1), but to perform a semiclassical approximation along the lines of [4]. We therefore make the ansatz $\Psi_k(\alpha, f_k) = e^{iS(\alpha, f_k)}$ and expand S in terms of powers of m_P^2:

$$S(\alpha, f_k) = m_P^2 S_0 + m_P^0 S_1 + m_P^{-2} S_2 + \dots \tag{3}$$

We then insert this ansatz into equation (1) and compare terms of equal power of m_P. At order $\mathcal{O}\left(m_P^2\right)$, we recover the Hamilton-Jacobi equation that describes the classical minisuperspace background on which the perturbations propagate. At the next order $\mathcal{O}\left(m_P^0\right)$, we define the wave functions $\psi_k^{(0)}(\alpha, f_k) \equiv \gamma(\alpha) e^{iS_1(\alpha, f_k)}$ and impose a condition on $\gamma(\alpha)$ in order to make it equal to the standard WKB prefactor. At this point, it is possible to introduce a time parameter t that arises from the minisuperspace background defined by the above-mentioned Hamilton-Jacobi equation:

$$\frac{\partial}{\partial t} := -e^{-3\alpha} \frac{\partial S_0}{\partial \alpha} \frac{\partial}{\partial \alpha}. \tag{4}$$

It follows that each of the wave functions $\psi_k^{(0)}(\alpha, f_k)$ obeys a Schrödinger equation with respect to the time parameter t:

$$i \frac{\partial}{\partial t} \psi_k^{(0)} = \mathcal{H}_k \psi_k^{(0)}. \tag{5}$$

Therefore, we can conclude that the order $\mathcal{O}\left(m_P^0\right)$ corresponds to the limit of quantum theory in an external background and we shall obtain the power spectrum of the scalar field perturbations at this order. In order to do so, we use the Gaussian ansatz $\psi_k^{(0)}(t, f_k) = \mathcal{N}_k^{(0)}(t) e^{-\frac{1}{2} \Omega_k^{(0)}(t) f_k^2}$ to solve equation (5). With the solution for $\Omega_k^{(0)}(t)$ evaluated at the time t_{exit}, when the respective mode exits the Hubble radius during the inflationary phase, we can calculate the density contrast $\delta_k(t)$ at the time t_{enter}, when the corresponding mode k reenters the Hubble radius during the radiation-dominated epoch, according to

$$\delta_k(t_{\text{enter}}) \propto \left| \frac{d}{dt} \mathfrak{Re}\left[\Omega_k^{(0)}(t) \right]^{-1/2} \right|_{t_{\text{exit}}}. \tag{6}$$

Consequently, it follows that $\delta_k(t_{\text{enter}}) \propto k^{-3/2}$. This implies an approximately scale-invariant power spectrum

$$\mathcal{P}^{(0)}(k) \propto k^3 \left| \delta_k(t_{\text{enter}}) \right|^2 \propto H^4 \left| \dot{\phi}(t) \right|_{t_{\text{exit}}}^{-2} \approx \text{const.}, \tag{7}$$

as it is found as a standard result for the simplest models of inflation.

4. The quantum-gravitational modification of the power spectrum

Since we intend to obtain quantum-gravitational correction terms to the power spectrum calculated above, we have to go one order higher in our semiclassical approximation, i.e. to the order $\mathcal{O}\left(m_\mathrm{p}^{-2}\right)$. Here, we define the wave functions $\psi_k^{(1)}(\alpha, f_k)$ and find that these obey the following quantum-gravitationally corrected Schrödinger equation

$$\mathrm{i}\frac{\partial}{\partial t}\psi_k^{(1)} = \mathcal{H}_k\psi_k^{(1)} - \frac{\mathrm{e}^{3\alpha}}{2m_\mathrm{p}^2\psi_k^{(0)}}\left[\frac{(\mathcal{H}_k)^2}{V}\psi_k^{(0)} + \mathrm{i}\frac{\partial}{\partial t}\left(\frac{\mathcal{H}_k}{V}\right)\psi_k^{(0)}\right]\psi_k^{(1)}, \qquad (8)$$

where $V := \mathrm{e}^{6\alpha}H^2$. It turns out that the first term in the brackets gives the dominant contribution and that the second term is negligible with respect to the first one, therefore we shall neglect the second term from now on.

We now have to find an approximate solution to equation (8) in order to calculate how the quantum-gravitational correction term modifies the power spectrum derived above. For this, we assume that one can accommodate the correction by a modified Gaussian ansatz of the form

$$\psi_k^{(1)}(t, f_k) = \left(\mathcal{N}_k^{(0)}(t) + \frac{1}{m_\mathrm{p}^2}\mathcal{N}_k^{(1)}(t)\right)\exp\left[-\frac{1}{2}\left(\Omega_k^{(0)}(t) + \frac{1}{m_\mathrm{p}^2}\Omega_k^{(1)}(t)\right)f_k^2\right]. \qquad (9)$$

We also demand that the quantum-gravitational correction vanishes for late times, which leads to the boundary condition $\Omega_k^{(1)}(t) \to 0$ as $t \to \infty$. It follows that the quantum-gravitational modification can be incorporated into a correction term $C(k)$ that relates the uncorrected power spectrum $\mathcal{P}^{(0)}$ to the quantum-gravitationally corrected one $\mathcal{P}^{(1)}$ according to

$$\mathcal{P}^{(1)}(k) = \mathcal{P}^{(0)}(k)\,C(k). \qquad (10)$$

The correction term $C(k)$ then takes the form

$$C(k) = 1 + \frac{\delta^\pm}{k^3}\frac{H^2}{m_\mathrm{p}^2} + \frac{1}{k^6}\mathcal{O}\left(\frac{H^4}{m_\mathrm{p}^4}\right), \qquad (11)$$

where the prefactor $\delta^\pm$ is a real constant. In the first article on this topic [1], we reported that the prefactor takes the value $\delta^- = -247.68$. However, in a more recent article [5], it was found that the implementation of the boundary condition can be subtly changed, which yields the alternative value $\delta^+ = 179.09$. But while there is a sign change, the order of magnitude of the quantum-gravitational correction stays the same.

5. Conclusion and outlook

The quantum-gravitational correction term in equation (8) leads to an explicit breaking of the approximate scale invariance of the uncorrected power spectrum and to a modification of the power at large scales, i.e. small values of k. However, due to the appearance of the ratio $(H/m_\mathrm{P})^2$ in equation (11), the quantum-gravitational modification could only become sizable if the inflationary Hubble parameter H approached the Planck scale. The tensor-to-scalar ratio r derived from measurements of the CMB anisotropies yields an upper bound on the energy scale of inflation and therefore on H, which is about $H \lesssim 4 \times 10^{-6}\, m_\mathrm{P} \approx 10^{14}\,\mathrm{GeV}$. We therefore see that even in this limiting case, we have to deal with an extremely small quantum-gravitational effect. Additionally, since cosmic variance limits the measurement accuracy at large scales fundamentally, such a small effect will not be able to be seen with more precise measurements of the CMB anisotropies by PLANCK or further future satellite missions. A more elaborate discussion on the observational bounds is presented in [6].

Notwithstanding of this pessimistic outlook, our analysis can be used to derive an upper bound on the Hubble parameter independently of the bound deduced from the tensor-to-scalar ratio. Based on CMB observations [7–9], the power spectrum deviates from a scale-invariant spectrum by less than 5 %. Hence, by using the assumption that our correction term $C(k)$ has to be within the limits $0.95 \lesssim C(k) \lesssim 1.05$ for $k \sim 1$, we can derive the bound $H \lesssim 1.5 \times 10^{-2}\, m_\mathrm{P} \approx 4 \times 10^{17}\,\mathrm{GeV}$, which is, however, weaker than the above-mentioned tensor-to-scalar ratio bound.

Acknowledgements

The author thanks Claus Kiefer for useful remarks and acknowledges support from the Bonn-Cologne Graduate School of Physics and Astronomy.

References

[1]   C. Kiefer, M. Krämer, "Quantum Gravitational Contributions to the Cosmic Microwave Background Anisotropy Spectrum", *Phys. Rev. Lett.* **108** (2012), 021301.

[2]   C. Kiefer, M. Krämer, "Can effects of quantum gravity be observed in the cosmic microwave background?", *Int. J. Mod. Phys.* **D21** (2012), 1241001.

[3]   J. J. Halliwell, S. W. Hawking, "The Origin of Structure in the Universe", *Phys. Rev.* **D31** (1985), 1777.

[4]   C. Kiefer, T. P. Singh, "Quantum gravitational corrections to the functional Schrödinger equation", *Phys. Rev.* **D44** (1991), 1067.

[5] D. Bini et al., "On the modification of the cosmic microwave background aniso-tropy spectrum from canonical quantum gravity" (2013), arXiv:1303.0531.

[6] G. Calcagni, "Observational effects from quantum cosmology" (2012), arXiv:1209.0473.

[7] E. Komatsu et al., "Seven-Year Wilkinson Microwave Anisotropy Probe (WMAP) Observations: Cosmological Interpretation", *Astrophys. J. Suppl. Ser.* **192** (2011), 18.

[8] G. Hinshaw et al., "Nine-Year Wilkinson Microwave Anisotropy Probe (WMAP) Observations: Cosmological Parameter Results" (2012), arXiv:1212.5226.

[9] P. A. R. Ade et al., "Planck 2013 results. XVI. Cosmological parameters" (2013), arXiv:1303.5076.

Gravitational-thermodynamic instabilities of isothermal spheres in dS and AdS

Zacharias Roupas

Institute of Nuclear and Particle Physics, N.C.S.R. Demokritos,
GR-15310 Athens, Greece

E-mail: roupas@inp.demokritos.gr

Abstract

Thermodynamical stability of fluid spheres is studied in the presence of a cosmological constant, both in the Newtonian limit, as well as in General Relativity. In all cases, an increase of the cosmological constant tends to stabilize the system, making asymptotically de Sitter space more thermodynamically stable than anti-de Sitter at the purely classical level. In addition, in the Newtonian case reentrant phase transitions are observed for a positive cosmological constant, due to its repelling property in this case. In General Relativity is studied the case of radiation, for which is found that the critical radius, at which an instability sets in, is always bigger than the black hole radius of the system and furthermore, at some value of the cosmological constant this critical radius hits at the cosmological horizon.

1. Introduction

Thermodynamics and stability analysis of self-gravitating gas in Newtonian Gravity is an old subject that dates back at least to the works of Antonov [1] and Lynden-Bell & Wood [2]. A classic review on the subject is written by Padmanabhan [3], while others include Refs. [4–7], as well. In General Relativity, thermodynamical stability of fluid spheres is studied by the author in Ref. [8], in which, this talk is, partially, based. The thermodynamics of self-gravitating gas is different from the ordinary thermodynamics, where the interactions are assumed to take place only between neighbouring bodies. Striking differences are the facts that entropy and energy are not extensive quantities, stable configurations with negative specific heat do exist and the thermodynamic ensembles are *not equivalent*. This non-equivalence of ensembles is not regarding the equilibria, but only the stability of these equilibria. That is, in gravitational thermodynamics the various thermodynamic ensembles have different stability properties.

We are interested in the effect of a cosmological constant to the stability of self-gravitating gas both in the Newtonian limit as well as in General Relativity. We shall not be restricted in the assumed current value of the cosmological constant, but we shall consider any arbitrary value. One can think physically about that, as regarding different possible Universes or as regarding the current Universe in possible different epochs in the past for cosmological models with a time varying cosmological term (e.g. decaying vacuum approaches [9–11]). Many results presented in this talk appear in Refs. [12–14], where is studied the thermodynamical stability in Newtonian Gravity with a cosmological constant in the microcanonical [12] and canonical [13, 14] ensembles.

Let us review briefly the results in Newtonian Gravity in the case with no cosmological constant, which for convenience we shall simply call 'flat case'. In the original formulation [1, 2], the self-gravitating gas, consisting of dust particles or stars, is assumed to be bound by a spherical, non-insulating and perfectly reflecting wall, so that it is in the microcanonical ensemble. For this system, global entropy maxima do not exist as Antonov have proved. There exist only local entropy maxima that correspond to metastable equilibria. These exist only for $ER > -0.335GM^2$ where R is the radius of the sphere and E, M the energy and mass of the gas, respectively. Thus, there is a minimum energy down to which equilibria do exist and strangely a *maximum* radius up to which equilibria do exist for a system with negative energy (like the vast majority of gravitating systems in nature). That is, for bigger radii than this maximum radius the system has no equilibria and collapses! This is usually called 'gravothermal catastrophe'. In addition there is another 'weaker' instability, associated with equilibria, that are unstable. Whether the system lies on an unstable or stable equilibrium (in the case $ER > -0.335GM^2$) depends on the density contrast $\log(\rho_0/\rho_R)$, where ρ_0, ρ_R are the densities at the center and the edge, respectively, or equivalently on the temperature. In the canonical ensemble, the system is assumed to lie in a heat bath and the walls are considered to be insulating. For this, it is known [2, 7] that equilibria do exist only for $TR > 0.40GM$ where T is the temperature of the system. Thus, there is a minimum temperature and a minimum radius down to which equilibria do exist. Note, that this behaviour is more similar to the general relativistic case, where a self-gravitating system collapses for small and not big radii. There is also a weak instability associated with unstable equilibria. Just like before, whether the system lies on an unstable or stable equilibrium (in the case $TR > 0.40GM$) depends on the density contrast $\log(\rho_0/\rho_R)$ or equivalently on the energy in this case.

We find [12] that in the microcanonical ensemble the presence of a positive cosmological constant tends to stabilize the system while a negative cosmological constant tends to destabilize it. We call a positive cosmological constant 'dS case', while a negative one

'AdS case'. An increase in the cosmological constant Λ causes an increase to the critical radius (up to which equilibria do exist) call it R_A, a decrease in the critical energy (down to which equilibria do exist) and a decrease in the critical density contrast that triggers the weak instability. In dS case, a reentrant behaviour is observed, since there appears a second critical radius we call R_{IA}, bigger than R_A, associated with Λ. At this radius the equilibria are restored. In the canonical ensemble [13, 14], for an increasing Λ the critical radius (down to which equilibria do exist) is decreasing, the critical temperature (down to which equilibria do exist) is decreasing and the critical density contrast is decreasing as well. In dS case, a reentrant phase transition occurs, since there appears a second lower critical temperature where equilibria are restored.

In General Relativity, the walls are not needed in order for the system to come into an equilibrium. A maximum entropy principle has been studied in Refs. [15–17]. In the case of relativistic radiation it has been proven by Sorkin, Wald and Zhang (SWZ) [16] for the specific case of radiation only, that the relativistic equation of hydrostatic equilibrium, namely the Tolman-Oppenheimer-Volkov equation (TOV) can be derived by the extremization of entropy and that the microcanonical thermodynamical stability coincides with the linear dynamical stability. Clarifying and generalizing an argument of Gao [17], we show [8] that TOV can be deduced thermodynamically by either the microcanonical or the canonical ensemble for any equation of state. Most importantly, we show the equivalence of microcanonical thermodynamical stability with linear dynamical stability for a general equation of state in General Relativity.

To study the effect of the cosmological constant on the microcanonical thermodynamical stability of fluid spheres in General Relativity we consider the fluid to be bounded by spherical non-insulating walls. For simplicity we consider only the case of radiation. We find that the critical radius down to which equilibria do exist is decreasing with increasing Λ. This radius is bigger than the black hole radius of the system for any value of Λ. At some Λ value the critical radius hits the cosmological horizon. Any sphere with bigger Λ is stable since matter outside the cosmological horizon cannot interact with matter inside. Finally, dS tends to stabilize the system while AdS to destabilize it, since the critical ratio M/R is increasing with increasing Λ.

2. Newtonian Gravity

In the presence of the cosmological constant the Poisson equation becomes [13]:

$$\nabla^2 \phi = 4\pi G \rho - 8\pi G \rho_\Lambda, \tag{1}$$

where ρ_Λ is the mass density associated with the cosmological constant Λ given by $\rho_\Lambda = \Lambda c^2/8\pi G$. The cosmological constant acts as a radial harmonic force: $F = (8\pi G/3)\rho_\Lambda r \hat{e}_r$, that is repulsive in dS case ($\rho_\Lambda > 0$) and attractive in AdS case ($\rho_\Lambda < 0$).

We consider a self-gravitating gas of N particles with unity mass inside a spherical shell. We restrict only to spherical symmetric configurations and work in the mean field approximation, where it is used the Boltzmann entropy

$$S = -k \int f \ln f \, d^3r \, d^3p, \qquad (2)$$

that is defined in terms of the one body distribution function $f(r, v)$. In the canonical ensemble is used the Helmholtz free energy $F = E - TS$ or equivalently we work with the Massieu function $J = -F/T$ that gives $J = S - \frac{1}{T}E$. The maximization of S with constant E and the maximization of J with constant T (and constant M in both cases), with respect to perturbations $\delta\rho$, are the same [13] to first order in $\delta\rho$, i.e. give the same equilibria, described by the Maxwell-Boltzmann distribution function

$$f = \left(\frac{\beta}{2\pi}\right)^{\frac{3}{2}} \rho(r) e^{-\frac{1}{2}\beta v^2}, \qquad \rho(r) = \rho_0 e^{-\beta(\phi - \phi(0))}, \qquad (3)$$

where ϕ satisfies equation (1). Although the two ensembles have the same equilibria, the second variations of S and J are different. A positive $\delta^2 S$ or $\delta^2 J$ for an equilibrium would signify that this equilibrium is unstable in the microcanonical or the canonical ensemble, respectively, since only local maxima of S or J correspond to stable equilibria.

Introducing the dimensionless variables $y = \beta(\phi - \phi(0))$, $x = r\sqrt{4\pi G\rho_0\beta}$ and $\lambda = 2\rho_\Lambda/\rho_0$, and using equation (3), equation (1) becomes:

$$\frac{1}{x^2}\frac{d}{dx}\left(x^2\frac{d}{dx}y\right) = e^{-y} - \lambda, \qquad (4)$$

called the Emden-Λ equation. Let us call $z = R\sqrt{4\pi G\rho_0\beta}$ the value of x at R. In order to generate the series of equilibria needed to study the stability of the system, the Emden-Λ equation has to be solved with initial conditions $y(0) = y'(0) = 0$, keeping M constant and for various values of the parameters ρ_Λ, β, ρ_0. This is a rather complicated problem, since, unlike $\Lambda = 0$ case, while solving for various z, mass is not automatically preserved, because of the mass scale $M_\Lambda = \rho_\Lambda \frac{4}{3}\pi R^3$ that Λ introduces. A suitable λ value has to be chosen at each z. We define the dimensionless mass

$$m \equiv \frac{M}{2M_\Lambda} = \frac{3}{8\pi}\frac{M}{\rho_\Lambda R^3} = \frac{\bar\rho}{2\rho_\Lambda}, \qquad (5)$$

where $\bar\rho$ is the mean density of matter. Calling $z = R\sqrt{4\pi G\rho_0\beta}$ the value of x at R, m can also be written as $m = 3B/\lambda z^2$, where

$$B = \frac{GM\beta}{R} \qquad (6)$$

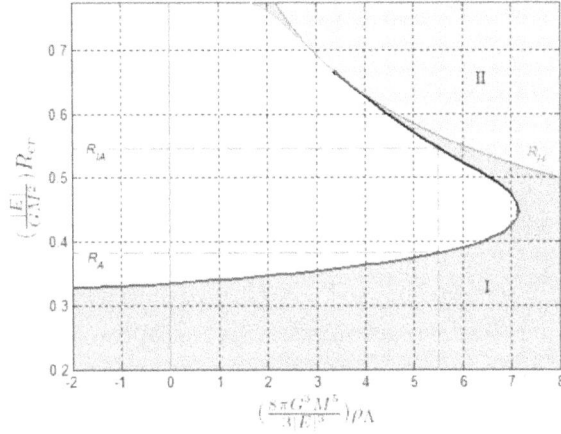

(a) Newtonian in the microcanonical ensemble

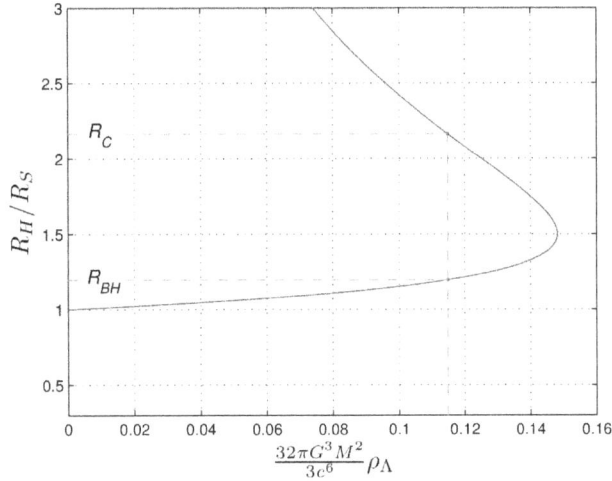

(b) Schwarzschild-dS

Figure 1.: On the top, the critical radius, in the Newtonian self-gravitating gas, versus ρ_Λ for fixed E, M in the microcanonical ensemble. There exist no equilibria in the unshaded region. With R_{H} is denoted the radius of the homogeneous solution. On the bottom, the horizons of the Schwarzschild-dS space versus the cosmological constant for a fixed mass. The horizon radius R_{H} is measured in units of the Schwarzschild radius $R_{\mathrm{S}} = \frac{2GM}{c^2}$. For a given cosmological constant there are two horizons: the black hole horizon R_{BH} and the cosmological horizon R_{C}. The similarity with Figure 1(a) is striking! Both figures appeared originally in Ref. [12].

is the dimensionless inverse temperature. It can be calculated by integrating the Emden-Λ equation, to get:

$$B(z) = zy'(z) + \frac{1}{3}\lambda z^2 . \tag{7}$$

We developed an algorithm to solve equation (4) for various values of λ, z keeping m fixed. From equation (5) it is clear that solving for various fixed m can be interpreted as solving for various ρ_Λ and/or R for a fixed M.

The dimensionless energy is defined as $Q = \frac{ER}{GM^2}$. It can be calculated by use of the virial theorem $2K + U_N - 2U_\Lambda = 3PV$, where K, U_N, U_Λ, P and V are the kinetic energy, the Newtonian potential energy, the cosmological potential energy, the pressure and the volume, respectively. The dimensionless energy Q is found [12, 13] to be:

$$Q(z) = \frac{z^2 e^{-y}}{B^2} - \frac{3}{2B} - \frac{\lambda}{2B^2 z} \int e^{-y} x^4 dx . \tag{8}$$

According to Poincaré's theorem [18, 19], the maximum of $Q(z)$ in a series of equilibria z, is the turning point of stability in the microcanonical ensemble, while the maximum of $B(z)$ is the turning point of stability in the canonical ensemble. At this equilibrium points a stable branch of equilibria becomes unstable. Whether the system lies on a stable or unstable equilibrium depends then on the density contrast $\log(\rho_0/\rho_R)$. The very same points mark the marginal conditions for which equilibrium states to exist, since for any Q or B, respectively, above these points there are no equilibria at all.

In dS case there are found *multiple* series of equilibria. For some of them the density is increasing towards the edge and others have periodic condensations, like a core-halo structure with one or even more haloes. In addition there appears a homogeneous solution at the radius $R_H = (3M/8\pi\rho_\Lambda)^{1/3}$, that is the equivalent to Einstein's static Universe in the Newtonian limit. This solution is stable for temperatures higher than the critical value $T_h = GM/6.73R_H$. In Figure 1(a), in the region I are solutions with $R < R_H$ and monotonically decreasing density towards the edge of the sphere, in region II there are solutions with $R > R_H$ and monotonically increasing density as well as solutions with periodic condensations. In the small gray area there are solutions with $R < R_H$ and periodic condensations. In the same figure, it is evident a reentrant behaviour in the critical radius. At the radius R_{IA}, equilibria are restored. A mysterious similarity with the relativistic Schwarzschild-dS space, that is an empty space, is obvious in Figure 1(b). It seems that the reentrant behaviour we observe is the Newtonian analogue of the two horizons in Schwarzschild-dS space. In this analogy, the Newtonian radius $R_N = GM^2/|E|$ is the analogue of the Schwarzschild radius $R_S = 2GM/c^2$. Note that Figure 1(a) is produced by rather complex numerical calculations, while in Figure 1(b) are drawn just the real roots of a simple third order algebraic polynomial.

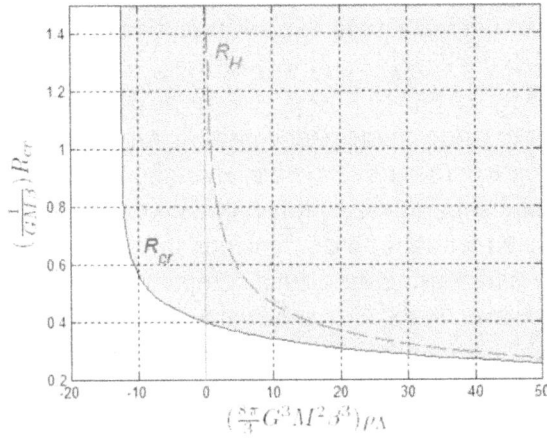

(a) Critical radius in the canonical ensemble

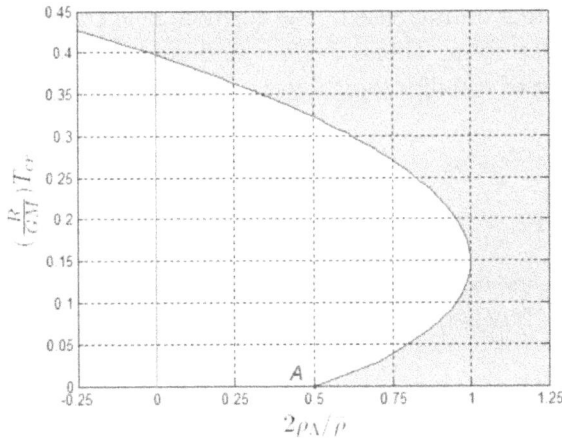

(b) Critical temperature in the canonical ensemble

Figure 2.: On the top, the critical radius versus ρ_Λ for fixed β, M in the canonical ensemble. There exist no equilibria in the unshaded region. With R_H is denoted the radius of the homogeneous solution. On the bottom, the critical temperature versus ρ_Λ for fixed M, R in the canonical ensemble, where $\bar{\rho}$ is the mean density of matter. In the unshaded region there exist no equilibria. This behaviour indicates a reentrant phase transition. Both figures appeared originally in Ref. [13].

In the canonical ensemble the system becomes unstable for radii *smaller* than a critical radius. In contrast to the microcanonical ensemble, this is a much more intuitively expected behaviour, similar to the one in General Relativity. The reason of the complete inversion of the region of instability in the two ensembles, lies on the the fact that in the microcanonical ensemble, the pressure gradient will increase during a compression of the system because of the increase in temperature, while in the canonical ensemble, the heat bath will absorb the energy and the pressure gradient will decrease, becoming unable to balance Gravity. In Figure 2(a) is shown the critical radius in the canonical ensemble with respect to the cosmological constant. An increase in the cosmological constant causes a decrease to the critical radius, enlarging the region of stability. In AdS case, we see that beyond some Λ value, it is impossible to retain any equilibrium. In Figure 2(b) we observe a reentrant phase transition in the critical temperature. As the cosmological constant is increasing the critical temperature, down to which equilibria exist, is decreasing. However, in dS, above the value $\rho_\Lambda^{\text{marg.}} = \bar{\rho}/4$ a further decrease in the temperature restores the equilibria beyond some second critical temperature. That happens because of the repelling character of positive Λ, which for low temperature can balance gravity. The value $\rho_\Lambda^{\text{marg.}}$ is this value of the cosmological constant that can balance gravity point by point at zero temperature, i.e. the point A in Figure 2(b) is a static *mechanical* equilibrium. The value $\rho_\Lambda^{\text{marg.}}$ is analytically calculated in [12].

Finally, we stress out that an increase of the cosmological constant tends to stabilize the system in both ensembles. This is evident in Figures 1(a), 2(a), 2(b) where one can see that the region of stability is increasing as ρ_Λ increases.

3. General Relativity

In General Relativity, a spherically symmetric, static perfect fluid obeys the Tolman-Oppenheimer-Volkov (TOV) equation:

$$p' = -\left(\frac{p}{c^2} + \rho\right)\left(\frac{Gm(r)}{r^2} + 4\pi G \frac{p}{c^2} r - \frac{8\pi G}{3}\rho_\Lambda r\right)\left(1 - \frac{2Gm(r)}{rc^2} - \frac{8\pi G}{3c^2}\rho_\Lambda r^2\right)^{-1},$$

(9)

with $m' = 4\pi r^2 \rho(r)$, where p is the pressure of the fluid and $m(r)$ the total mass-energy (including the one of the gravitational field) included inside r. This equation expresses the hydrostatic equilibrium in General Relativity and is derived by the Einstein's equations. In Ref. [8], we showed following [16, 17] that it can also be derived by the extremization of entropy in the microcanonical ensemble or the extremization of free energy in the canonical ensemble. Most importantly, we showed that the second variation of entropy gives the same condition for stability with radial variations to first order in Einstein's equations for a general equation of state. Thus, is evident the equivalence between microcanonical thermodynamical stability and linear dynamical stability. We observed in the previous chapter that the Newtonian canonical ensemble, behaves similar to a relativistic system. We will see in the followings, even more evidence, that the microcanonical ensemble in General Relativity becomes the canonical ensemble in the Newtonian limit, hinting at the presence of an implicit heat bath in General Relativity.

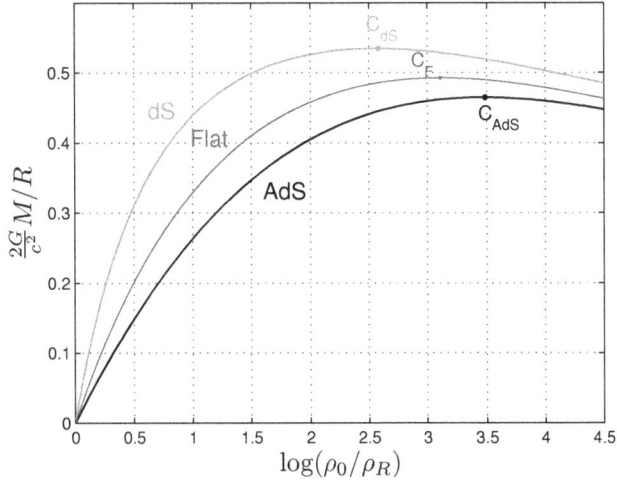

(a) Ratio M/R for $q = 1/3$

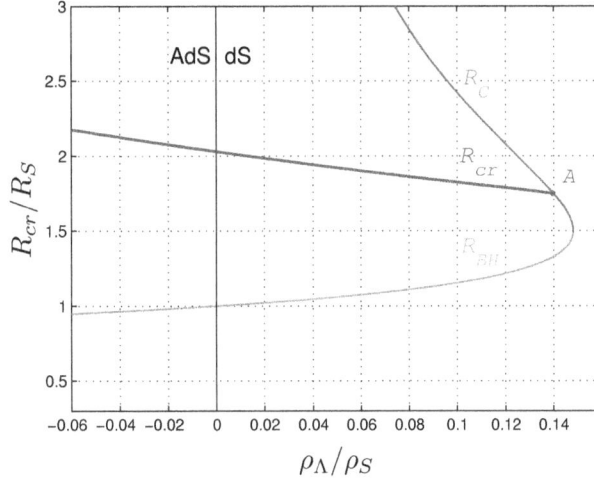

(b) Critical radius for $q = 1/3$

Figure 3.: On the top, the series of equilibria expressed as the ratio M/R with respect to the density contrast $\log(\rho_0/\rho_R)$ for radiation ($q = 1/3$), for asymptotically flat, de Sitter and anti-de Sitter cases. The points C_i are turning points of stability, i.e. at their left side the equilibria are stable, while at their right side, unstable. At each case above C_i there are no equilibria at all. On the bottom, the critical radius R_{cr} with respect to the cosmological constant ρ_Λ in Schwarzschild units $R_S = 2GM/c^2$, $\rho_S = 3M/4\pi R_S^3$ for a fixed mass M. The spheres of radius $R < R_{cr}$ are strongly thermodynamically unstable, i.e. there exist no equilibria at all. At point A, the critical radius hits the cosmological horizon. Figure 3(b) appeared originally in Ref. [8].

For a linear equation of state

$$p = q\rho c^2 \tag{10}$$

and introducing the dimensionless variables

$$\rho = \rho_0 e^{-y}, \quad x = r\sqrt{4\pi G\rho_0 \frac{q+1}{qc^2}}, \quad \lambda = \frac{2\rho_\Lambda}{\rho_0},$$

$$\mu(x) = \frac{1}{4\pi\rho_0}\left(4\pi G\rho_0 \frac{q+1}{qc^2}\right)^{\frac{3}{2}} m(r), \quad M = m(R), \quad z = x(R), \tag{11}$$

the TOV equation (9) becomes

$$\frac{dy}{dx} = \left(\frac{\mu}{x^2} + qxe^{-y} - \frac{\lambda}{3}x\right)\left(1 - 2\frac{q}{q+1}\frac{\mu}{x} - \frac{\lambda}{3}\frac{q}{q+1}x^2\right)^{-1},$$

$$\frac{d\mu}{dx} = x^2 e^{-y}. \tag{12}$$

The Emden-Λ equation (4) is the limit of the above equation for dust matter $q \to 0$. Just like in the Newtonian case we define the dimensionless mass $\tilde{m} \equiv \frac{M}{2M_\Lambda} = \frac{3}{8\pi}\frac{M}{\rho_\Lambda R^3} = \frac{3\mu}{\lambda z^3}$ and develop a computer code to solve equation (12) keeping $\tilde{m}$ fixed. The dimensionless energy becomes now

$$Q \equiv \frac{2GM}{Rc^2} = \frac{2\mu}{z}\frac{q}{q+1} = \frac{2q}{q+1}\frac{zy'(z) - qz^2 e^{-y(z)} - \frac{\lambda}{3}z^2\left(\frac{q}{q+1}zy'(z) + 1\right)}{2\frac{q}{q+1}zy'(z) + 1}. \tag{13}$$

Weinberg (see p. 305 in Ref. [20]) has proven that the maxima of M coincide with the maxima of the baryon number N in a series of equilibria, for systems with constant chemical composition and constant entropy per baryon, like systems with a linear equation of state. The maxima of M just like the Newtonian case are turning points of stability for the microcanonical ensemble, in accordance with the Poincaré's theorem. So, in our case (linear equation of state) the maxima of N are turning points, as well. We use the dimensionless baryon number

$$B(z) \equiv \frac{N}{N_*} = \frac{1}{z^{\frac{3q+1}{q+1}}}\int_0^z x^2 e^{-\frac{y}{q+1}}\left(1 - 2\frac{q}{q+1}\frac{\mu}{x} - \frac{\lambda}{3}\frac{q}{q+1}x^2\right)^{-\frac{1}{2}} dx,$$

$$N_* = 4\pi R^3\left(\frac{1}{4\pi GKR^2}\frac{q^2c^4}{q+1}\right)^{\frac{1}{q+1}}, \tag{14}$$

where K is the constant that enters in the polytropic equation $p = Kn^{q+1}$ with n the baryon number density. The Newtonian limit $q \to 0$ of B is the quantity

$$B \xrightarrow{q\to 0} \frac{GM\beta}{R},$$

that is exactly the dimensionless temperature we used in the Newtonian limit, that controls canonical stability. Since as we have seen B controls microcanonical stability in relativity, we deduce that the Newtonian limit of the microcanonical ensemble is the canonical ensemble! This conclusion is supported by even stronger evidence; (a) in [8] is proven that the condition for stability in the microcanonical ensemble in General Relativity, becomes the condition for stability in the Newtonian canonical ensemble for non-relativistic dust matter (b) the Newtonian canonical ensemble behaves qualitatively similar to relativistic systems.

Regarding the effect of the cosmological constant on the stability of the system, we focus in the case of radiation, i.e. $q = 1/3$. In Figure 3(b) we see how the critical radius, down to which equilibria do exist, is changing with respect to the cosmological constant. The system becomes unstable long before it reaches its black hole radius for any value of the cosmological constant. At some value, it equals the cosmological horizon. Any bigger sphere can be regarded stable, since matter outside the horizon cannot interact with matter inside.

In Figure 3(a) is plotted the ratio M/R of with respect to the density contrast for various equilibria. The maximum point is a turning point of stability and in the same time the marginal point for which equilibria do exist. Since this point goes to bigger values as Λ increases we conclude that an increase in the cosmological constant tends to stabilize the system just like in the Newtonian limit.

Acknowledgements

Many of the results used in this talk were elaborated jointly with Minos Axenides and George Georgiou and appear analytically elsewhere.

References

[1] V. Antonov, "Most Probable Phase Distribution in Spherical Star Systems and Conditions for Its Existence", *Vest. Leningrad Univ.* **7** (1962), 135.

[2] D. Lynden-Bell, R. Wood, "The Gravo-thermal Catastrophe in Isothermal Spheres and the onset of Red-giant Structure for stellar systems", *MNRAS* **138** (1968), 495.

[3] T. Padmanabhan, "Statistical Mechanics of Gravitating Systems", *Phys. Rep.* **188** (1990), 285.

[4] T. Katz, "Thermodynamics of Self-Gravitating Systems", *Found. Phys.* **33** (2003), 223.

[5] H. de Vega, N. G. Sanchez, "Statistical mechanics of the selfgravitating gas. 1. thermodynamic limit and phase diagram", *Nucl. Phys.* **B625** (2002), 409.

[6] H. de Vega, N. G. Sanchez, "Statistical mechanics of the selfgravitating gas. 2. Local physical magnitudes and fractal structures", *Nucl. Phys.* **B625** (2002), 460.

[7] P.-H. Chavanis, "Gravitational instability of finite isothermal spheres", *A& A* **381** (2001), 340.

[8] Z. Roupas, "Thermodynamical instabilities of perfect fluid spheres in General Relativity" (2013), arXiv:1301.3686.

[9] I. Waga, "Decaying Vacuum Flat Cosmological Models: expressions for some observable quantities and their properties", *Astrophys. J.* **414** (1993), 436.

[10] R. Woodard, N. Tsamis, "Quantum Gravity slows inflation", *Nucl. Phys.* **B474** (1996), 235.

[11] A. Polyakov, "Infrared instability of the de Sitter space " (2012), arXiv:1209.4135.

[12] M. Axenides, G. Georgiou, Z. Roupas, "Gravothermal Catastrophe with a Cosmological Constant", *Phys. Rev.* **D86** (2012), 104005.

[13] M. Axenides, G. Georgiou, Z. Roupas, "Gravitational instabilities of isothermal spheres in the presence of a cosmological constant", *Nucl. Phys.* **B871** (2013), 21.

[14] M. Axenides, G. Georgiou, Z. Roupas, "Gravothermal instability with a cosmological constant in the canonical ensemble", *J. Phys.: Conf. Ser.* **410** (2013), 012130.

[15] W. Cocke, "A maximum entropy principle in general relativity and the stability of fluid spheres", *Ann. Inst. Henri Poncaré* **2** (1965), 283.

[16] R. Sorkin, R. Wald, Z. Zhang, "Entropy of self-gravitating radiation", *Gen. Rel. Grav.* **13** (1981), 1127.

[17] S. Gao, "A general maximum entropy principle for self-gravitating perfect fluid", *Phys. Rev.* **D84** (2011), 104023.

[18] H. Poincaré, "Sur l'équilibre d'une masse fluide animée d'un mouvement de rotation", *Acta. Math.* **7** (1885), 259.

[19] T. Katz, "On the number of unstable modes of an equilibrium", *MNRAS* **183** (1978), 765.

[20] S. Weinberg, "Gravitation and Cosmology: principles and applications of the general theory of relativity" (1972).

Geometry of almost-product Lorentzian manifolds and relativistic observer

Andrzej Borowiec[1] and **Aneta Wojnar**[2]

Instytut Fizyki Teoretycznej, Uniwersytet Wrocławski,
pl. M. Borna 9, 50-204 Wrocław, Poland

E-mail: [1]borow@ift.uni.wroc.pl, [2]aneta@ift.uni.wroc.pl

Abstract
The notion of relativistic observer is confronted with Naveira's classification of (pseudo-)Riemannian almost-product structures on spacetime manifolds. Some physical properties and their geometrical counterparts are shortly discussed.

1. Introduction

In Einstein's General Relativity, a gravitational interaction is represented by a metric with Lorentzian signature $(-, +, +, +)$ living on a (curved) four-dimensional spacetime manifold and satisfying Einstein's field equations. An observer is an independent notion and, according to a nowadays point of view, can be identified with an arrow of time. More precisely, the observer is determined by a timelike normalized (local) vector field on spacetime. We can also think of it as the collection of its integral curves, considered as world lines (also known as the congruence of world lines of point observers) of some continuous material object (e.g. relativistic fluid). From a mathematical perspective, it provides a one-dimensional (timelike) foliation. It appears that a pair, the metric and the vector field, determines a differential-geometric structure which is called an almost-product structure. From a physical perspective, a relativistic observer is tautologically defined as a field of his own four-velocities. Having chosen an observer, one can define relativistic observables, i.e. relative measurable quantities. They include the relative (three-)velocity of another observer or test particles (see e.g. [1],[2], [3]), as well as Noether conserved currents in diffeomorphism covariant field theories [4]. The well-known splitting of the electromagnetic field into measurable electric and magnetic components is also relative to the observer. In the more traditional approach to General Relativity, the measurable quantities are related to coordinates. In fact, given a coordinate system, one can associate to it a (local) observer, indicated by a time variable. However the notion adopted here is more general, coordinate-free and can be globalized.

In the presented note we provide the correspondence between Naveira's classes of a pseudo-Riemannian manifold [5] implemented by the observer and its physical characteristics as introduced in [6]. The paper is organized as follows. In section 2 we introduce the notation and basic notions. In section 3 we shortly recall Gil-Medrano's theorem [7], which provides a differential geometric interpretation for Naveira's classes. The advantages of the almost-product structures in physics are discussed in section 4 (see also [8] in this context). They extend the possible characteristics for a given observer on a Lorentzian manifold. Finally, we provide a few illustrative examples in section 5.

2. Preliminaries and definitions

Let M and TM denote respectively an n-dimensional smooth manifold and its tangent bundle. A k-dimensional ($k < n$) tangent distribution (k-distribution in short) is a map D which associates a k-dimensional subspace $D_p \subset T_pM$ to the point $p \in M$:

$$D : p \rightarrow D_p \subset T_pM. \tag{1}$$

D can be also considered as a subbundle of TM. Locally, one can say that a k-distribution is generated by a set of k linearly independent vector fields iff in every point p their values span the k-dimensional subspace D_p, i.e.

$$D_p = \text{span}\{X_1(p),\ldots,X_k(p)\}. \tag{2}$$

In this case we shall write $X_i \in \Gamma(D)$, where $\Gamma(D)$ stands for a submodule of cross sections of the subbundle $D \subset TM$. An embedded submanifold $N \subset M$ is called an integral manifold of the distribution D if $T_pN = D_p$ in every point $p \in N$. We say that D is involutive if, for each pair of local vector fields (X,Y) belonging to D, their Lie bracket $[X,Y]$ is also a vector field from D. The distribution D is completely integrable if for each point $p \in M$ there exists an integral manifold N of the distribution D passing through p such that the dimension of N is equal to the dimension of D. It turns out that every involutive distribution is completely integrable (local Frobenius theorem). Every smooth 1-dimensional distribution is integrable. The integrability of a distribution is closely related to the notion of foliation. We have the following (global) Frobenius theorem:

Theorem 1 *Let D be an involutive k-dimensional tangent distribution on a smooth manifold M. The collection of all maximal connected integral manifolds of D forms a foliation of M.*

The proof of the theorem and the precise definition of a foliation can be found in [9]. Roughly speaking, a foliation is a collection of submanifolds N_i such that each submanifold proceeds smoothly into another one. They do not cross each other. Particularly, a class of globally hyperbolic spacetimes $M = T \times \Sigma$, where T is an open interval in the real line $\mathbb{R}$ and Σ is a three-manifold, serve as a typical example of global foliation [10]. Let us recall [11, 12] that an almost-product structure on M is determined by a field of endomorphisms of TM, i.e. a $(1,1)$ tensor field P on M, such that $P^2 = I$ ($I =$Identity). In this case, at any point $p \in M$, one can consider two subspaces of $T_p M$ corresponding respectively to two eigenvalues ± 1 of P. It defines two complementary distributions on M, i.e. $TM = D^+ \oplus D^-$. Moreover, if M is equipped with a (pseudo -)Riemannian metric g such that

$$g(PX, PY) = g(X, Y); \; X, Y \in \Gamma(TM), \tag{3}$$

then both distributions are mutually orthogonal. In this case, P is called a (pseudo -) Riemannian almost-product structure. It is to be noticed that some modified gravity models admit almost-product structures as solutions [13].

3. Geometric characterization of distributions on (pseudo-)Riemannian manifolds

Let D be a distribution on (M, g) and $D^\perp$ the distribution orthogonal to D. At every point $p \in M$, we have then $T_p M = D_p \oplus D_p^\perp$ [1]. Thus we can uniquely define a $(1,1)$ tensor field P such that $P^2 = I$, $P_{|D} = 1$, $P_{|D^\perp} = -1$. It is clear that P becomes automatically a (pseudo -)Riemaniann almost-product structure. One has (see [7]):

Definition 1 *The distribution D is called geodesic, minimal or umbilical if and only if D has property D_1, D_2 or D_3 respectively, where:*

- $D_1 \iff (\nabla_A P)A = 0$,

- $D_2 \iff \alpha(X) = 0$,

- $D_3 \iff g((\nabla_A P)B, X) + g((\nabla_B P)A, X) = \frac{2}{k} g(A, B)\alpha(X)$,

where $X \in \Gamma(D)^\perp$; $A, B \in \Gamma(D)$. Here $\{e_a\}_{a=1}^k$ ($k = \dim D$) is a local orthonormal frame of D and $\alpha(X) = \sum_{a=1}^k g((\nabla_{e_a} P)e_a, X)$.

It implies that a distribution has the property D_1 if and only if it has the properties D_2 and D_3. Their meanings in the case of integrability are explained below.

---

[1] The case of null distributions is more complicated and should be discussed separately, see e.g. [14].

45

Theorem 2 *(O. Gil-Medrano) A foliation D is called totally geodesic, minimal or totally umbilical if and only if D has the property F_1, F_2 or F_3 respectively, where*

$$F_i \iff F + D_i, \quad i = 1, 2, 3 \tag{4}$$

and

$$F \iff (\nabla_A P)B = (\nabla_B P)A \ \forall A, B \in \Gamma(D). \tag{5}$$

The proof of this theorem can be found in [7]. It is easy to see that the property F is equivalent to Frobenius' theorem, i.e. a distribution D with this property is a maximal foliation. The theorem says that, in principle, one deals with three special types of foliations:

(F_1) Totally geodesic foliation: it means that every geodesic of an arbitrary integral submanifold N (the leaf of foliation), if considered together with the induced metric (the first fundamental form), is at the same time geodesic of the total manifold M. Moreover, it is equivalent to the statement that the second fundamental form of N (i.e. extrinsic curvature) vanishes. In other words, the extrinsic curvature measures the failure of a geodesic of the manifold N to be a geodesic of M.

(F_2) Minimal foliation: If there is a surface with the smallest possible value of the area bounded by a certain curve, that surface is called a minimal surface. The condition for a distribution to be a minimal distribution is that the trace of the second fundamental form vanishes. The trace of the extrinsic curvature is also called mean curvature, that is, the average of the principal curvatures. Examples of minimal surfaces in $\mathbb{R}^3$ are the catenoid and the helicoid.

(F_3) Umbilical foliation: We recall that an umbilical manifold is a manifold for which all points are umbilical points. Umbilical points, in turn, are locally spherical: every tangent vector at such point is a principal direction and all principal curvatures are equal [15]. For example, a sphere is an umbilical manifold. In the case of integral submanifolds, the second fundamental form has to be proportional to the induced metric.

4. Almost-product structure related to a spacetime observer

In the present section we are going to apply the formalism presented above to the special case of a relativistic observer on a spacetime manifold. These new tools will be used at the end of the section for a final classification.

From now on (M, g) denotes a four-dimensional manifold (spacetime) equipped with Lorentzian signature metric $g_{\alpha\beta}$. An observer is represented by a timelike vector field u^α which, according to our sign convention $(-, +, +, +)$, is normalized to

$$u^\alpha u_\alpha = -1. \tag{6}$$

Strictly speaking, the normalization condition (6) prevents the existence of critical points and one can deal with a one-dimensional (timelike) distribution instead. Such a distribution is always integrable and provides a foliation with world-lines as leaves. Each leaf can then be parameterized by arc length (proper time), making u^α a four-velocity field. This implies that the only nontrivial question one can ask about a one-dimensional distribution is weather it is geodesic or not (see below the tables).

Because of this, one should concentrate on its orthogonal (transverse) completion D. This is a spacelike three-dimensional distribution with Euclidean signature. These two distributions provide a $3+1$ (orthogonal) decomposition of the tangent bundle $TM = D \oplus D^\perp$, with the one-dimensional timelike distribution denoted as $D^\perp$. It is easy to find out that the corresponding three-dimensional projection tensor has the form:

$$h^\alpha_\beta = \delta^\alpha_\beta + u^\alpha u_\beta, \tag{7}$$

which, due to (6), implies $h^\alpha_\rho h^\rho_\beta = h^\alpha_\beta$. We would like to stress that in what follows we shall always use the original metric $g_{\alpha\beta}$ for lowering and rising indices. Thus covariant and contravariant components of tensors can be used exchangeably. For example, the second-rank symmetric tensor

$$h_{\alpha\beta} = g_{\alpha\beta} + u_\alpha u_\beta, \tag{8}$$

plays the role of induced Euclidean metric on the distribution D. When D is integrable, then (8) is the first fundamental form (i.e. induced metric) on each leaf. The corresponding foliation by spacelike hypersurfaces has the physical meaning of clock synchronization and divides the spacetime into equal-time pieces identified as three-dimensional spaces. One should mention that the integrability of D is always required in the case of $3+1$ splitting which is necessary for the Hamiltonian formalism of General Relativity (see e.g. [16]).

More generally, to any tensor $A^{\alpha\cdots}_{\beta\cdots}$ living in the spacetime one can assign its projected three-dimensional counterpart

$$\tilde{A}^{\alpha\cdots}_{\beta\cdots} = h^\alpha_\mu h^\nu_\beta \cdots A^{\mu\cdots}_{\nu\cdots} \tag{9}$$

According to widely spread ideas (see e.g. [6, 10, 17]), only projected three-dimensional tensors are good candidates for measurable relativistic observables. Obviously, such quantities are relative, i.e. observer dependent. For example, for an anti-symmetric covariant two-tensor $F_{\alpha\beta} = -F_{\beta\alpha}$ (two-form), which under the closeness condition ($dF = 0$) can be interpreted as an electromagnetic field, one gets

$$F_{\alpha\beta} = H_{\alpha\beta} + u_\alpha E_\beta - u_\beta E_\alpha, \tag{10}$$

where $H_{\alpha\beta} = \tilde{F}_{\alpha\beta} = h^\mu_\alpha h^\nu_\beta F_{\mu\nu}$ and $E_\alpha = u^\mu h^\nu_\alpha F_{\mu\nu}$ are measurable electric and magnetic components.

Before proceeding further, let us answer the question of when the one-dimensional foliation spanned by u is totally geodesic. This can be easily done by studying the auto-parallel (geodesic) equation

$$u^\beta u_{\alpha;\beta} = 0, \tag{11}$$

where $u_{\alpha;\beta} = \nabla_\beta u_\alpha$ denotes the Levi-Civita covariant derivative of u. Thus introducing the acceleration vector $\dot{u}_\alpha = u^\beta u_{\alpha;\beta}$ one can conclude that the vanishing of $\dot{u}^\alpha$ is equivalent to the geodesic equation (11). One should notice that $\dot{u}_\alpha$ is, in fact, a three-vector, since $u^\alpha \dot{u}_\alpha = 0$.

In general, one can decompose the space components of the two-tensor $u_{\alpha;\sigma}$ into irreducible parts with respect to the three-dimensional orthogonal group:

$$\tilde{u}_{\alpha;\beta} = h^\sigma_\beta u_{\alpha;\sigma} = \omega_{\alpha\beta} + \sigma_{\alpha\beta} + \frac{1}{3}\Theta h_{\alpha\beta}, \tag{12}$$

where $\omega_{\alpha\beta}$ denotes its antisymmetric part, $\sigma_{\alpha\beta}$ is the traceless symmetric component and finally Θ stands for the trace. This is a kinematical decomposition.[2] Using (7) we shall obtain [6, 19]:

$$u_{\alpha;\beta} = \omega_{\alpha\beta} + \sigma_{\alpha\beta} + \frac{1}{3}\Theta h_{\alpha\beta} - \dot{u}_\alpha u_\beta. \tag{13}$$

There is a well-known interpretation of the observer in terms of relativistic hydrodynamics, treating it as a flow of material points constituting a (perfect) fluid (continuous medium), with the world lines being particle trajectories: one line of the flow passes through every point x^α of a certain spacelike (possibly bounded) region in spacetime. Accordingly, the tensor $u_{\alpha;\sigma}$ determines the rate of change in the position of one point with respect to the other one in the material [20].

Keeping in mind this fluid analogy, each irreducible component of the projected tensor $u_{\alpha;\beta}$ admits a physical interpretation which is contained in self-explanatory and intuitive names (for more detailed explanations see e.g. [6, 20, 21]). In a more explicit form, one has to take into account the following three-dimensional quantities:

$$\omega_{\alpha\beta} = u_{[\alpha;\beta]} + \dot{u}_{[\alpha} u_{\beta]} \qquad \text{is the rotation tensor,} \tag{14}$$

$$\sigma_{\alpha\beta} = u_{(\alpha;\beta)} + \dot{u}_{(\alpha} u_{\beta)} - \frac{1}{3}\Theta h_{\alpha\beta} \qquad \text{is the shear tensor,} \tag{15}$$

$$\Theta = u^\alpha_{;\alpha} \qquad \text{is the expansion scalar,} \tag{16}$$

$$\dot{u}^\alpha = u^\beta u^\alpha_{;\beta} \qquad \text{is the acceleration vector.} \tag{17}$$

It is more convenient to use the scalars

$$\dot{u} \equiv \left(\dot{u}_\alpha \dot{u}^\alpha\right)^{\frac{1}{2}}, \quad \omega \equiv \left(\frac{1}{2}\omega_{\alpha\beta}\omega^{\alpha\beta}\right)^{\frac{1}{2}}, \quad \sigma \equiv \left(\frac{1}{2}\sigma_{\alpha\beta}\sigma^{\alpha\beta}\right)^{\frac{1}{2}}. \tag{18}$$

---

[2]The dynamical equation is known as Raychaudhuri equation (see e.g. [17, 18]).

These are non-negative and vanish at the same time as their corresponding tensors. An observer is rotation-free, shear-free or expansion-free when $\omega = 0$, $\sigma = 0$ or $\Theta = 0$ respectively. If all quantities vanish, then the observer is called rigid. It is worth mentioning that the observer (four-velocity) field u^α can be used to construct the energy momentum tensor of an ideal (incompressible) fluid,

$$T_{\alpha\beta} = (p + \rho)u_\alpha u_\beta + p g_{\alpha\beta}, \tag{19}$$

where the matter density ρ and the pressure p are internal fluid parameters determining its thermodynamical behavior. The same energy momentum tensor treated on the right-hand side of Einstein's equations as the source of the gravitational field influences the metric. This suggests possible relationships between metric and fluid observer, which are an interesting subject for future research (see e.g. [18, 22]).

Now we are ready to classify all almost-product structures related to relativistic observers in gravitational spacetimes. As we have already mentioned, the tensor h^α_β projects on a three-dimensional subspace while $-u^\alpha u_\beta = \delta^\alpha_\beta - h^\alpha_\beta$ projects on the one-dimensional complementary distribution spanned by u^α. It turns out that the difference:

$$P^\alpha_\beta = h^\alpha_\beta - (-u^\alpha u^\beta) = \delta^\alpha_\beta + 2u^\alpha u_\beta. \tag{20}$$

represents an almost-product structure compatible with the metric g.[3] Now the almost-product structure (20) can be used to encode the observer u^α.

Since the issue of one-dimensional distributions have already been solved, we should concentrate on the three-dimensional one. One has 4 conditions to be imposed on P (see Definition 1 and Theorem 2). The umbilical case D_3, after some manipulations, produces

$$u_{(\alpha;\beta)} + \dot{u}_{(\alpha}u_{\beta)} = \frac{1}{3}\sum_{i=1}^{3} e_i^\beta u_{\alpha;\beta} e_i^\alpha. \tag{21}$$

Here $\{e_i\}_{i=1}^3$ denotes a local orthonormal frame of D. We notice that the sum of the right hand side is a three-dimensional trace of the tensor $u_{\alpha;\beta}$, thus D_3 is equivalent to the vanishing of the shear tensor.

The condition D_2 (a minimal distribution) leads to

$$\sum_{i=1}^{3} e_i^\beta u_{\alpha;\beta} e_i^\alpha = 0, \tag{22}$$

which is equivalent to the vanishing a scalar of expansion.

For a geodesic distribution (D_1), one obtains

$$u_{(\alpha;\beta)} + \dot{u}_{(\alpha}u_{\beta)} = 0, \tag{23}$$

---

[3]It is easy to see that P satisfies the conditions $P^2 = I$, as well as (3).

which is equivalent to the vanishing of both characteristics: shear and expansion. In the free falling case ($\dot{u} = 0$) the condition (23) denotes that the normalized timelike vector u^α is a Killing vector for the metric $g_{\mu\nu}$.

Similarly, one can show that the integrability condition (F) reduces to

$$u_{[\alpha;\beta]} + \dot{u}_{[\alpha} u_{\beta]} = 0, \tag{24}$$

which means vanishing of a rotation.

The final results are presented in the Table 1 and Table 2. The first one concerns accelerated $\dot{u} \neq 0$ observers with all possibilities for the three-dimensional distribution taken into account. Similarly, the second table concerns free-falling observers ($\dot{u} = 0$). The bracket $(1,3)$ in the first column indicates that the first symbol is for the one-dimensional distribution and the second for the three-dimensional one. The tables also contain the physical interpretations for each class of almost-product Lorentzian manifolds.

Class (1,3)	Accelerated observers $\dot{u} \neq 0$	Physical meaning	3-distribution
$(F,-)$	$u_{\alpha;\beta} = \sigma_{\alpha\beta} + \omega_{\alpha\beta} + \frac{1}{3}\Theta h_{\alpha\beta} - \dot{u}_\alpha u_\beta$	——	non-integrable **distribution**
(F,D_2)	$\Theta = 0 \Rightarrow u_{\alpha;\beta} = \sigma_{\alpha\beta} + \omega_{\alpha\beta} - \dot{u}_\alpha u_\beta$	expansion-free	minimal
(F,D_3)	$\sigma = 0 \Rightarrow u_{\alpha;\beta} = \omega_{\alpha\beta} + \frac{1}{3}\Theta h_{\alpha\beta} - \dot{u}_\alpha u_\beta$	shear-free	umbilical
(F,D_1)	$\Theta = \sigma = 0 \Rightarrow u_{\alpha;\beta} = \omega_{\alpha\beta} - \dot{u}_\alpha u_\beta$	shear-free & expansion-free	geodesic
(F,F)	$\omega = 0 \Rightarrow u_{\alpha;\beta} = \sigma_{\alpha\beta} + \frac{1}{3}\Theta h_{\alpha\beta} - \dot{u}_\alpha u_\beta$	rotation-free	**foliation**
(F,F_2)	$\omega = \Theta = 0 \Rightarrow u_{\alpha;\beta} = \sigma_{\alpha\beta} - \dot{u}_\alpha u_\beta$	rotation-free & expansion-free	minimal
(F,F_3)	$\omega = \sigma = 0 \Rightarrow u_{\alpha;\beta} = \frac{1}{3}\Theta h_{\alpha\beta} - \dot{u}_\alpha u_\beta$	rotation-free & shear-free	totally umbilical
(F,F_1)	$u_{\alpha;\beta} = -\dot{u}_\alpha u_\beta$	rigid	totally geodesic

Table 1.: Accelerated observers

5. Illustrative examples

5.1. Minkowski spacetime

The most extreme case in Naveira's classification is (F_1, F_1) class, i.e. both distributions are totally geodesic foliations. In Minkowski spacetime the metric is flat in the

Class (1,3)	Geodesic (free falling) observers $\dot{u} = 0$	Physical meaning	3-distribution
$(F_1, -)$	$u_{\alpha;\beta} = u_{[\alpha;\beta]} + \sigma_{\alpha\beta} + \frac{1}{3}\Theta h_{\alpha\beta}$	geodesic	**non-integrable distribution**
(F_1, D_2)	$\Theta = 0 \Rightarrow u_{\alpha;\beta} = u_{[\alpha;\beta]} + \sigma_{\alpha\beta}$	geodesic expansion-free	minimal
(F_1, D_3)	$\sigma = 0 \Rightarrow u_{\alpha;\beta} = u_{[\alpha;\beta]} + \frac{1}{3}\Theta h_{\alpha\beta}$	geodesic shear-free	umbilical
(F_1, D_1)	$\sigma = \Theta = 0 \Rightarrow u_{\alpha;\beta} = u_{[\alpha;\beta]}$	geodesic shear-free & expansion-free	geodesic
(F_1, F)	$\omega = 0 \Rightarrow u_{\alpha;\beta} = \sigma_{\alpha\beta} + \frac{1}{3}\Theta h_{\alpha\beta}$	geodesic rotation-free	**foliation**
(F_1, F_2)	$\omega = \Theta = 0, \Rightarrow u_{\alpha;\beta} = \sigma_{\alpha\beta}$	geodesic rotation-free & expansion-free	minimal
(F_1, F_3)	$\omega = \sigma = 0 \Rightarrow u_{\alpha;\beta} = \frac{1}{3}\Theta h_{\alpha\beta}$	geodesic rotation-free & shear-free	totally umbilical
(F_1, F_1)	$u_{\alpha;\beta} = 0$	geodesic rigid	totally geodesic

Table 2.: Free-falling observers

Cartesian coordinate system ($\eta_{\mu\nu} = \text{diag}(-1, 1, 1, 1)$), so one can replace the covariant derivatives with partial ones. Then the (F_1, F_1) case becomes just $u_{\alpha,\beta} = 0$. There exists a solution in the form of a constant vector. In fact, any constant timelike vector field can be changed by a linear transformation of coordinates (i.e. Lorentz transformation) into

$$u^\alpha = [1, 0, 0, 0].\qquad(25)$$

Such a vector field is a canonical inertial observer in Minkowski spacetime. A less restrictive class is (F, F_1), which implies that the observer should accelerate. An example of such an observer is Rindler's:

$$u^\alpha = \left[\frac{x}{x^2 - t^2}, \frac{t}{x^2 - t^2}, 0, 0\right],\qquad(26)$$

for whom only a part of Minkowski space is available. The only non-vanishing characteristic is the acceleration

$$\dot{u} = (x^2 - t^2)^{-1/2},\qquad(27)$$

which is constant along each trajectory. Again, by introducing adapted (Rindler's) coordinates one can simplify expressions.

Let us consider the rotating observer in the (x, y) plane,

$$u^\alpha = \left[\sqrt{2}, \frac{-y}{\sqrt{x^2+y^2}}, \frac{x}{\sqrt{x^2+y^2}}, 0 \right], \tag{28}$$

belonging to the class (F, D_2) with the following characteristics:

$$\Theta = 0, \tag{29}$$

$$\dot{u} = (x^2 + y^2)^{-1/2}, \tag{30}$$

$$\omega = \frac{\sqrt{2}}{2}(x^2 + y^2)^{-1/2}, \tag{31}$$

$$\sigma = \frac{\sqrt{2}}{2}(x^2 + y^2)^{-1/2}, \tag{32}$$

constant along particles trajectories. The last example for Minkowski spacetime (in the spherical coordinates $g_{\mu\nu} = \text{diag}(-1, 1, r^2, r^2 \sin^2 \theta)$) is the observer

$$u^\alpha = \left[\frac{r}{\sqrt{r^2 - t^2}}, \frac{t}{\sqrt{r^2 - t^2}}, 0, 0 \right]. \tag{33}$$

It turns out that observer (33) has the following characteristics:

$$\dot{u} = (r^2 - t^2)^{-1/2}, \tag{34}$$

$$\omega = 0, \tag{35}$$

$$\Theta = \frac{2t}{r}(r^2 - t^2)^{-1/2}, \tag{36}$$

$$\sigma = \frac{\sqrt{3}t}{3r}(r^2 - t^2)^{-1/2}, \tag{37}$$

and belongs to the class (F, F).

5.2. Schwarzschild spacetime

There is no observer belonging to the class (F_1, F_1) in Schwarzschild spacetime. It should satisfy the sixteen equations $u_{\alpha;\beta} = 0$, which turn out to be inconsistent. Let us consider the observer

$$u^\alpha = \left[\left(1 - \frac{2M}{r}\right)^{-\frac{1}{2}}, 0, 0, 0 \right]. \tag{38}$$

The only non-vanishing characteristic is the acceleration $\dot{u} = \frac{M}{r^2}(1 - \frac{2M}{r})^{-1/2}$, which implies that observer (38) belongs to the class (F, F_1).

The geodesic observer [17] is of the form

$$u^\alpha = \left[\left(1 - \frac{3M}{r}\right)^{-1/2}, 0, \sqrt{\frac{M}{r^2(r-3M)}}, 0 \right]. \tag{39}$$

It shows a singular expansion at the north and south poles:

$$\Theta = \sqrt{\frac{M}{r^2(r-3M)}} \cot\theta. \tag{40}$$

The rotation and shear scalars are

$$\omega = \frac{1}{4}\sqrt{\frac{M}{r^3}}\left(\frac{1-6M/r}{1-3M/r}\right), \tag{41}$$

$$\sigma = \sqrt{\frac{27M\sin^2\theta(r-2M)^2 + 16Mr(r-3M)\cos^2\theta}{48r^3(r-3M)^2\sin^2\theta}}. \tag{42}$$

It belongs to the $(F, -)$ class, so for this observer, as well as for (28) of Minkowski spacetime, there is no three-dimensional orthogonal distribution providing a foliation of the spacetime manifold (M, g) i.e. there are no three-dimensional equal-time subspaces relative to these observers.

Acknowledgements

A.W. would like to thank the Organizers for warm hospitality during the Conference. The calculations have been partially performed in Maxima.

References

[1]   D. Bini, P. Carini, R. Jantzen, "Relative observer kinematics in general relativity", *Class. Quant. Grav.* **12** (1995), 2549.

[2]   Z. Oziewicz, "Relativity groupoid insted of relativity group", *Int. J. Geom. Meth. Mod. Phys.* **4** (2007), 739.

[3]   A. Ungar, "Einstein's Special Relativity: The Hyperbolic Geometric Viewpoint" (2013), eprint: 1302.6961.

[4]   L. Fatibene et al., "ADM Pseudotensors, Conserved Quantities and Covariant Conservation Laws in General Relativity", *Ann. Phys.* **327** (2012), 1593.

[5]   A. Naveira, "A classification of Riemannian almost product manifolds", *Rend. Mat.* **3** (1983), 577.

[6] J. Ehlers, "Contributions to the relativistic mechanics of continuous media", *Gen. Rel. Grav.* **25**.12 (1993), 1225.

[7] O. Gil-Medrano, "Geometric properties of some classes of Riemannian almost-product manifolds", *Rend. Circ. Mat. Palermo* **32**.3 (1983), 315.

[8] B. Coll, J. Ferrando, "Almost-product structures in relativity", in *Recent Developments in Gravitation*, vol. 1, World Scientific (Singapore), 1990, p. 338.

[9] J. Lee, "Introduction to smooth manifolds", Springer Verlag (Berlin / Heidelberg / New York), 2000.

[10] S. Hawking, G. Ellis, "The large scale structure of space-time geometry", Cambridge University Press (Cambridge), 1973.

[11] A. Gray, "Pseudo-Riemannian almost product manifolds and submersions", *Math. Mech.* **16** (1967), 715.

[12] K. Yano, M. Kon, "Structures on manifolds", Vol. 3, World Scientific (New Jersey), 1984.

[13] A. Borowiec et al., "Almost-complex and almost-product Einstein manifolds from a variational principle", *J. Math. Phys.* **40** (1999), 3446.

[14] A. Bejancu, H. Farran, "Foliations and Geometric Structures", Vol. 580, Springer (Berlin / Heidelberg / New York), 2006.

[15] M. Spivak, "A Comprehensive Introduction to Differential Geometry", Vol. 4, Publish or Perish, Inc. (Houston), 1999.

[16] E. Gourgoulhon, "3+1 formalism and bases of numerical relativity" (2007), arXiv:gr-qc/0703035.

[17] E. Poisson, "A relativist's toolkit", Cambridge University Press (Cambridge), 2004.

[18] R. Slobodeanu, "Shear-free perfect fluids with linear equation of state" (2013), arXiv:1301.5508.

[19] C. Eckart, "The Thermodynamics of Irreversible Processes. III. Relativistic Theory of the Simple Fluid", *Phys. Rev.* **58** (1940), 919.

[20] J. Plebański, A. Krasiński, "An introduction to general relativity and cosmology", Cambridge University Press (Cambridge), 2006.

[21] M. Demiański, "Astrofizyka relatywistyczna", Wydawnictwo Naukowe PWN (Warszawa), 1991.

[22] M. Carrera, D. Giulini, "On the generalization of McVittie's model for an inhomogeneity in a cosmological spacetime", *Phys. Rev.* **D81** (2010), 043521.

supplemented with a mass term for the metric perturbations $h_{\mu\nu}$. The action is given by:

$$S_{FP} = \int d^4x \left\{ h^{\rho\sigma} \mathcal{E}_{\rho\sigma}{}^{\mu\nu} h_{\mu\nu} - \frac{1}{2} m^2 \left(h_{\mu\nu} h^{\mu\nu} - h^2 \right) \right\}, \tag{1}$$

where h is the trace of $h_{\mu\nu}$ and $h^{\rho\sigma} \mathcal{E}_{\rho\sigma}{}^{\mu\nu} h_{\mu\nu}$ is the linearised Einstein tensor. It contains different contractions of two derivatives in such a way that the whole expression is invariant under linearised diffeomorphisms $\delta h_{\mu\nu} = \partial_{(\mu} \xi_{\nu)}$. The mass term breaks this symmetry and as a consequence the theory propagates 5 degrees of freedom in four dimensions in contrast to the 2 d.o.f. of general relativity.

We will now study the limit where the graviton mass m goes to zero. At the end of the day, physical observables in this limit should be the same as those in general relativity. After coupling the action (1) to a spherically symmetric matter distribution sourced by a stress-energy tensor $T_{\mu\nu}$ we may take the limit $m \to 0$, calculate some observables and compare with linearised GR.

	General Relativity	FP with $m \to 0$
Light bending angle	$\frac{4G_N M}{b}$	$\frac{4G_N M}{b}$
Newtonian potential	$-\frac{G_N M}{r}$	$-\frac{4}{3} \frac{G_N M}{r}$

The above table shows that Fierz-Pauli theory in the limit of zero graviton mass produces a different Newtonian potential than GR. Since the light bending angle is not modified, absorbing the factor 4/3 into G_N does not resolve the problem. This is called the van Dam-Veltman-Zakharov (vDVZ) discontinuity [10, 11]. The infinitesimally small parameter m reproduces finite deviations in observables. This is in violation with the intuition that the physics should be continuous in the parameters of the theory.

The origin of the vDVZ discontinuity can be exposed when we decompose the 5 degrees of freedom of the massive graviton into 2 tensor (helicity ± 2), 2 vector (helicity ± 1) and 1 scalar (helicity 0) degree of freedom. This may be achieved by performing a Stueckelberg transformation of the field $h_{\mu\nu}$,

$$h_{\mu\nu} \to h'_{\mu\nu} + \partial_{(\mu} A_{\nu)} + 2 \partial_\mu \partial_\nu \phi + \eta_{\mu\nu} \phi. \tag{2}$$

After the transformation we see that the tensor ($h'_{\mu\nu}$) part describes linearised general relativity. In the $m \to 0$ limit the vector part decouples, but the scalar degree of freedom still couples to the trace of the stress tensor. So in fact, Fierz-Pauli theory in the $m \to 0$ limit describes linearised GR plus a scalar coupled to matter, which causes the inconsistency in the calculation of the Newtonian potential.

3. Non-linearities and the Boulware-Deser ghost

The trick to resolving the vDVZ-discontinuity is to take non-linear effects into account. As is well known, GR is a non-linear theory and at a distance scale set by the Schwarzschild radius, non-linear effects will start to play a role. In Fierz-Pauli theory this scale is set by the Vainshtein radius [12], which is given by

$$r_V = \left(\frac{G_N M}{m^4} \right)^{1/5}.$$ (3)

As the graviton mass is taken to zero, the Vainshtein radius will go to infinity. This entails that nowhere in this limit the linearised approximation may be trusted and full non-linear effects must be taken into account.

So a next logical approach may be to consider the full non-linear theory of general relativity, supplemented with a mass term for the fluctuations of the metric $h_{\mu\nu}$.

$$S = \int d^4x \sqrt{g} \left\{ R - \frac{1}{4} m^2 g^{(0)\mu\rho} g^{(0)\nu\sigma} (h_{\mu\nu} h_{\rho\sigma} - h_{\mu\rho} h_{\nu\sigma}) \right\},$$ (4)

where $g^{(0)}_{\mu\nu}$ is a fixed and invertible reference metric, which is needed to contract indices on $h_{\mu\nu}$. Usually this is taken to be the background metric.

We are now dealing with a different theory, and should investigate whether it propagates the right amount of degrees of freedom to describe a massive graviton. To this end one may decompose the metric $g_{\mu\nu}$ in ADM variables; the spatial metric g_{ij}, the lapse N and the shift N_i and investigate the phase space of the theory through the Hamiltonian. In general relativity one will find primary constraints that ensure that the canonical momenta of the lapse and shift vanish. These variables then appear in the Hamiltonian as Lagrangian multipliers for a set of secondary constraints. In d dimensions the counting of degrees of freedom proceeds as follows. The spatial component of the metric is a symmetric $d - 1$ matrix, so it has $\frac{1}{2}d(d-1)$ components. From this we subtract the d constraints enforced by the lapse and shift, leading to a total of $\frac{1}{2}d(d-3)$ degrees of freedom.

The addition of a mass-term in the action (4) will introduce terms quadratic in the lapse and shift functions, so that they don't play the role of Lagrange multipliers any more, but instead become auxiliary fields. As a result, the degrees of freedom propagated by the theory are given by the $\frac{1}{2}d(d-1)$ components of the spatial metric. In 4 dimensions, these are 6 degrees of freedom, 5 for the massive graviton and in addition there is an extra ghost-like scalar degree of freedom. This is the Boulware-Deser (BD) ghost [13]. Since the kinetic term for this scalar ghost carries the wrong sign, it is responsible for a repulsive "fifth force". It can be shown that this fifth force exactly cancels the attractive force caused by the scalar coupling to the trace of the stress-tensor which was responsible for the vDVZ-discontinuity. The presence of the ghost thus resolves the discontinuity in what is known as the Vainshtein mechanism. However, its presence also creates an unstable vacuum for this non-linear completion of Fierz-Pauli theory.

4. Removing the ghost

As we have seen in the last section, non-linear extensions of Fierz-Pauli theory contain a sixth degree of freedom which is a scalar ghost. Here we will discuss two recent approaches to massive gravity which are free of this BD-ghost: dRGT massive gravity and New Massive Gravity (NMG).

4.1. dRGT massive gravity

The origin of the ghost mode in (4) can be seen in the Stückelberg formulation of the theory. The massive graviton can be decomposed into helicity ± 2, helicity ± 1 and helicity 0 modes. The leading order interactions for the helicity 0 mode are suppressed by an energy scale $\Lambda_5 = (M_P m^4)^{1/5}$. The dynamics of this mode can then be studied in the decoupling limit [14]: $M_P \to \infty, m \to 0$ and Λ_5 fixed. This limit sends all operators suppressed by an energy scale higher than Λ_5 to infinity and thus decouples them from the theory. The kinetic term of the helicity 0 mode now receives four-derivative contributions which signals that the scalar sector propagates two degrees of freedom. One is a massless scalar field and the other the Boulware-Deser ghost, with a mass of order of the energy scale Λ_5. One may wonder whether this theory will still make sense as an effective theory of massive gravity at energies lower than this cut-off scale, so before the mass of the ghost mode becomes of order one. However, the distance scale set by the BD ghost coincides with the scale at which quantum corrections become important [14]. This scale is parametrically larger than the Vainshtein radius, implying that the quantum theory takes over before the classical non-linearities can restore the proper limit with general relativity. So in order to have a suitable effective theory of massive gravity, we can try to raise the cut-off scale by adding higher order interactions to the theory.

This is the approach of de Rham, Gabadaze and Tolley (dRGT) [15, 16]. They consider an action with an Einstein-Hilbert kinetic term and an interaction potential $U_n(g^{(0)}, h)$.

$$S_{dRGT} \propto \int d^4x \left\{ \sqrt{-g}R - \frac{1}{4}\sqrt{-g^{(0)}}m^2 \sum_{n=2}^{\infty} U_n(g^{(0)}, h) \right\} , \qquad (5)$$

with

$$U_2 = [h^2] - [h]^2,$$
$$U_3 = a_1[h^3] + a_2[h][h^2] + a_3[h]^3,$$
$$U_4 = b_1[h^4] + b_2[h][h^3] + b_3[h^2][h^2] + b_4[h^2][h]^2 + b_5[h]^4, \quad (6)$$
$$\vdots$$

where the square brackets denote taking the trace with the reference metric $[h] = g^{(0)\mu\nu}h_{\mu\nu}$. Notice that the first term, U_2, is just the Fierz-Pauli mass term. After introducing the Stueckelberg decomposition in this interaction potential, one can see that at every order in n higher derivative kinetic terms for the scalar field will appear. The trick is now to tune the coefficients (a_i for U_3, b_i for U_4, etc.) in such a way that the higher derivative scalar terms become a total derivative. This is possible at every order in n and fixes all but two coefficients in four dimensions. The tuning of these coefficients removes all scalar interaction terms suppressed by any scale $\Lambda < \Lambda_3 = (m^2 M_p)^{1/3}$. Then the new decoupling limit may be taken with Λ_3 fixed and the consequent effective Lagrangian has at most second order derivatives on the scalar field, signalling the absence of the ghost mode in the decoupling limit.

In order to study the ghost problem beyond the decoupling limit, the authors of [16] resum the interaction terms (6) into a fully non-linear theory of Massive Gravity involving elementary symmetric polynomials of the square root matrix $\sqrt{g^{-1}f}$, where $f_{\mu\nu}$ is the reference metric (see also [17]). This fully non-linear theory was shown to be ghost free in [18]. This is due to the fact that after an ADM decomposition of the metric, the shift function N remains a Lagrange multiplier and enforces a secondary constraint. The number of degrees of freedom is then $\frac{1}{2}d(d-1) - 1$ which correctly describes the degrees of freedom of a massive graviton in d dimensions.

4.2. New Massive Gravity

Another approach to a ghost free theory of massive gravity was developed by Bergshoeff, Hohm and Townsend [19, 20]. It does not involve adding a mass term and higher order interactions which break the diffeomorphism invariance of the theory, but instead one adds higher derivatives of the metric in an invariant way by adding squares of the curvature invariants $R^{\mu\nu}R_{\mu\nu}$ and R^2.

$$S_{\text{NMG}} = \frac{M_P}{2} \int d^3x \sqrt{-g} \left\{ \sigma R + \frac{1}{m^2} \left(R^{\mu\nu}R_{\mu\nu} - \frac{3}{8}R^2 \right) \right\}. \quad (7)$$

Note that this action is written in three dimensions and there is a sign parameter $\sigma = \pm 1$ in front of the Einstein-Hilbert term. This action has fourth order equations of motion for the metric. In general this would give rise to an Ostrogradski-type instability; the fourth order equations would give a solution corresponding to the massless graviton of GR and a new, massive, solution with the wrong sign in front of its kinetic term, i.e. a massive ghost graviton. This is the reason why New Massive Gravity is considered in three dimensions. Then the massless graviton does not propagate any degrees of freedom. If we take the sign parameter $\sigma = -1$, it will be the massless solution that carries the wrong sign in its kinetic term and the remaining theory will be a ghost-free theory of a massive graviton in three dimensions.

The advantage of New Massive Gravity (NMG) over adding an explicit mass term is that now the symmetries of GR are not broken. An obvious disadvantage is that this will only give a ghost-free theory in three dimensions, in higher dimensions either the massless or the massive graviton is a ghost, depending on the value of σ. Still gravity in three dimensions is an interesting subject in light of the gauge/gravity correspondence. Starting from the observations of Brown and Henneaux [21] that the asymptotic symmetry group of three dimensional gravity with a negative cosmological constant is two copies of the Virasoro algebra, three dimensional gravity in AdS space is now considered to be dual to a two dimensional conformal field theory (CFT). In NMG one can add a cosmological constant $\Lambda = -1/\ell^2$ and linearise around an Anti-de Sitter (AdS) space-time. The resulting Lagrangian is [22]

$$S = \frac{1}{2}\left(1 - \frac{1}{2m^2\ell^2}\right)h^{\mu\nu}\mathcal{E}_{\mu\nu}{}^{\rho\sigma}h_{\rho\sigma} + \frac{1}{2m^2}k^{\mu\nu}\mathcal{E}_{\mu\nu}{}^{\rho\sigma}h_{\rho\sigma} - \frac{1}{4m^2}\left(k^{\mu\nu}k_{\mu\nu} - k^2\right). \quad (8)$$

At a special, critical point in the parameter space of this theory, $m^2 = 1/2\ell^2$, the first term in this action vanishes and effectively the graviton mass goes to zero. At this point the linearised equations of motion are

$$\left(\Box - \frac{2}{\ell^2}\right)^2 h_{\mu\nu} = 0, \quad (9)$$

where $\Box = g^{(0)\mu\nu}\nabla_\mu\nabla_\nu$ is the d'Alembertian of the background metric. In addition to the massless graviton, which satisfies $\left(\Box - \frac{2}{\ell^2}\right)h^{(0)}_{\mu\nu} = 0$, there is a new solution given by

$$\left(\Box - \frac{2}{\ell^2}\right)^2 h^{\log}_{\mu\nu} = 0, \quad \text{but} \quad \left(\Box - \frac{2}{\ell^2}\right)h^{\log}_{\mu\nu} \propto h^{(0)}_{\mu\nu} \neq 0. \quad (10)$$

This solution is the so-called logarithmic mode [22, 23], since it does not satisfy the usual Brown-Henneaux boundary conditions in AdS space, but instead it falls off logarithmically towards the boundary. In this case, the Brown-Henneaux boundary conditions must be modified to account for this logarithmic falloff behaviour [24]. Now the asymptotic symmetry group is still two copies of the Virasoro algebra, but the logarithmic mode acts as a source for an operator with a non-zero Jordan cell. This means that the Hamiltonian of the dual field theory is not diagonalizable and the stress-tensor acquires a so-called logarithmic partner. This is characteristic of logarithmic CFTs (LCFT). The conjecture that New Massive Gravity at the critical point is dual to a logarithmic CFT has been checked by calculating the 2-point correlation functions through a process of holographic renormalization [25, 26] and through a calculation of the partition function in [27].

5. Parity Even Tricritical Gravity

The logarithmic CFT dual to critical NMG is of rank 2, meaning that the stress-energy tensor acquires one logarithmic partner and the Jordan cell of the theory has rank 2. It is possible to construct dual gravitational theories for LCFTs with higher rank by adding even higher derivative terms to the Lagrangian [28–31]. One such example is Parity Even Tricritical (PET) gravity, defined by the action [29]

$$S \sim \int d^3x \sqrt{-g} \left\{ \sigma R - 2\Lambda_0 + \alpha R^{\mu\nu} R_{\mu\nu} - \frac{1}{8} \left(3\alpha - \beta\Lambda \right) R^2 + \right.$$
$$\left. + \beta \left(\nabla_\rho R_{\mu\nu} \nabla^\rho R^{\mu\nu} - \frac{3}{8} \nabla_\mu R \nabla^\mu R \right) \right\}, \tag{11}$$

where the cosmological constant Λ is related to the bare parameters Λ_0, α and β as $\Lambda_0 = \sigma\Lambda - \frac{3}{4}\Lambda^3\beta + \frac{1}{4}\Lambda^2\alpha$. This theory has sixth order equations of motion and for arbitrary parameters it describes one massless graviton and two massive gravitons with masses

$$M_\pm^2 = \frac{\alpha}{2\beta} - \Lambda \pm \frac{1}{2\beta} \sqrt{10\beta^2\Lambda^2 - 6\beta\alpha\Lambda + 4\beta\sigma + \alpha^2}. \tag{12}$$

The parameter space of this theory has a special point when

$$\alpha = -4\frac{\sigma}{\Lambda} \quad \text{and} \quad \beta = -2\frac{\sigma}{\Lambda^2}, \tag{13}$$

since then $M_\pm^2 = 0$. This point corresponds to a tricritical point in parameter space, where both massive modes degenerate with the massless mode. In analogy to NMG, new solutions arise corresponding to a log mode and a $\log^2$ mode

$$(\Box + 2\Lambda)^3 h_{\mu\nu}^{\log^2} = 0, (\Box + 2\Lambda)^2 h_{\mu\nu}^{\log^2} \propto h_{\mu\nu}^{(0)}, \qquad (\Box + 2\Lambda) h_{\mu\nu}^{\log^2} \propto h_{\mu\nu}^{\log}. \tag{14}$$
$$(\Box + 2\Lambda)^2 h_{\mu\nu}^{\log} = 0, (\Box + 2\Lambda) h_{\mu\nu}^{\log} \propto h_{\mu\nu}^{(0)}. \tag{15}$$

The log and the $\log^2$ modes do not satisfy Brown-Henneaux boundary conditions in AdS$_3$, however, it is possible to relax the boundary conditions to allow for logarithmic and $\log^2$ falloff towards the boundary. In this case the operators dual to these modes are two logarithmic partners of the stress-tensor, and hence the dual theory is expected to be a rank-3 LCFT.

6. Summary and outlook

We have presented an historical overview of massive gravity and some recent achievements in removing the unstable Boulware-Deser scalar mode. The removal of this ghost mode could be achieved by either adding higher-order interactions in four dimensions or by considering higher derivative theories in three dimensions. The disadvantage of the first approach is that it explicitly relies on a fixed reference metric, while the latter approach only works in three dimensions. It is unclear whether phenomenologically viable four dimensional extensions of higher derivative gravities exist and this is an interesting area of research. For a recent attempt, see [32]. We have furthermore discussed a dual field theory interpretation of critically tuned higher-derivative massive gravities and provided some evidence that the dual theories are logarithmic CFTs.

An interesting generalisation of the dRGT models was presented in [33] where the fixed reference metric is promoted to a fully dynamical metric. These bimetric theories describe two dynamical metrics interacting through a potential which reproduces to the dRGT interaction potential when one of the two metrics is fixed to be the flat reference metric. In three dimensions another limit of the bimetric theory leads to New Massive Gravity [34]. In this limit one of the metrics will give rise to an auxiliary field whose equations of motion fix it to be proportional to the Schouten tensor. Back substitution of the auxiliary field into the Lagrangian then gives the NMG action. So it seems that the different approaches to massive gravity may be linked through theories with two dynamical metrics. Whether these bimetric theories are also free of ghost-like instabilities is subject of debate, compare for instance [35–39].

Acknowledgements

I would like to thank Eric Bergshoeff, Jan Rosseel, Sjoerd de Haan and Thomas Zojer for collaborating on some of the subjects presented here. Furthermore, thanks goes out to the organizers of the Barcelona Postgrad Encounters on Fundamental Physics for a wonderful conference and all the participants for creating a very stimulating and inspiring environment.

References

[1]   A. Einstein, "The Foundation of the General Theory of Relativity", *Ann. Phys. (Berlin)* **49** (1916), 769.

[2]   A. G. Riess, "Observational evidence from supernovae for an accelerating universe and a cosmological constant", *Astron. J.* **116** (1998), 1009.

[3]   S. Perlmutter, "Measurements of Omega and Lambda from 42 high redshift supernovae", *Astrophys. J.* **517** (1999), 565.

[4]   C. de Rham et al., "Cosmic Acceleration and the Helicity-0 Graviton", *Phys. Rev.* **D83** (2011), 103516.

[5]   S. N. Gupta, "Gravitation and Electromagnetism", *Phys. Rev.* **96** (1954), 1683.

[6]   S. Weinberg, "Photons and gravitons in perturbation theory: Derivation of Maxwell's and Einstein's equations", *Phys. Rev.* **138** (1965), B988.

[7]   S. Deser, "Selfinteraction and gauge invariance", *Gen. Rel. Grav.* **1** (1970), 9.

[8]   K. Hinterbichler, "Theoretical Aspects of Massive Gravity", *Rev. Mod. Phys.* **84** (2012), 671.

[9]   M. Fierz, W. Pauli, "On relativistic wave equations for particles of arbitrary spin in an electromagnetic field", *Proc. Roy. Soc. Lond.* **A173** (1939), 211.

[10]   H. van Dam, M. Veltman, "Massive and massless Yang-Mills and gravitational fields", *Nucl. Phys.* **B22** (1970), 397.

[11]   V. Zakharov, "Linearized gravitation theory and the graviton mass", *JETP Lett.* **12** (1970), 312.

[12]   A. Vainshtein, "To the problem of nonvanishing gravitation mass", *Phys. Lett.* **B39** (1972), 393.

[13]   D. Boulware, S. Deser, "Can gravitation have a finite range?", *Phys. Rev.* **D6** (1972), 3368.

[14]   N. Arkani-Hamed, H. Georgi, M. D. Schwartz, "Effective field theory for massive gravitons and gravity in theory space", *Ann. Phys.* **305** (2003), 96.

[15]   C. de Rham, G. Gabadadze, "Generalization of the Fierz-Pauli Action", *Phys. Rev.* **D82** (2010), 044020.

[16]   C. de Rham, G. Gabadadze, A. J. Tolley, "Resummation of Massive Gravity", *Phys. Rev. Lett.* **106** (2011), 231101.

[17]   S. Hassan, R. A. Rosen, "On Non-Linear Actions for Massive Gravity", *JHEP* **1107** (2011), 009.

[18]   S. Hassan, R. A. Rosen, "Resolving the Ghost Problem in non-Linear Massive Gravity", *Phys. Rev. Lett.* **108** (2012), 041101.

[19]   E. A. Bergshoeff, O. Hohm, P. K. Townsend, "Massive Gravity in Three Dimensions", *Phys. Rev. Lett.* **102** (2009), 201301.

[20]   E. A. Bergshoeff, O. Hohm, P. K. Townsend, "More on Massive 3D Gravity", *Phys. Rev.* **D79** (2009), 124042.

[21]   J. D. Brown, M. Henneaux, "Central Charges in the Canonical Realization of Asymptotic Symmetries: An Example from Three-Dimensional Gravity", *Commun. Math. Phys.* **104** (1986), 207.

[22] E. A. Bergshoeff et al., "Modes of Log Gravity", *Phys. Rev.* **D83** (2011), 104038.

[23] M. Alishahiha, R. Fareghbal, "D-Dimensional Log Gravity", *Phys. Rev.* **D83** (2011), 084052.

[24] Y. Liu, W. Sun, "Consistent Boundary Conditions for New Massive Gravity in AdS_3", *JHEP* **0905** (2009), 039.

[25] D. Grumiller, O. Hohm, "AdS(3)/LCFT(2): Correlators in New Massive Gravity", *Phys. Lett.* **B686** (2010), 264.

[26] M. Alishahiha, A. Naseh, "Holographic Renormalization of New Massive Gravity" (2010), arXiv:1005.1544.

[27] M. R. Gaberdiel, D. Grumiller, D. Vassilevich, "Graviton 1-loop partition function for 3-dimensional massive gravity", *JHEP* **1011** (2010), 094.

[28] E. A. Bergshoeff et al., "Unitary Truncations and Critical Gravity: a Toy Model", *JHEP* **1204** (2012), 134.

[29] E. A. Bergshoeff et al., "On Three-Dimensional Tricritical Gravity", *Phys. Rev.* **D86** (2012), 064037.

[30] T. Nutma, "Polycritical Gravities", *Phys. Rev.* **D85** (2012), 124040.

[31] A. Kleinschmidt, T. Nutma, A. Virmani, "On unitary subsectors of polycritical gravities" (2013), arXiv:1206.7095.

[32] E. A. Bergshoeff et al., "On 'New Massive' 4D Gravity", *JHEP* **1204** (2012), 070.

[33] S. Hassan, R. A. Rosen, "Bimetric Gravity from Ghost-free Massive Gravity", *JHEP* **1202** (2012), 126.

[34] M. F. Paulos, A. J. Tolley, "Massive Gravity Theories and limits of Ghost-free Bigravity models", *JHEP* **1209** (2012), 002.

[35] S. Hassan, R. A. Rosen, "Confirmation of the Secondary Constraint and Absence of Ghost in Massive Gravity and Bimetric Gravity", *JHEP* **1204** (2012), 123.

[36] K. Nomura, J. Soda, "When is Multimetric Gravity Ghost-free?", *Phys. Rev.* **D86** (2012), 084052.

[37] J. Kluson, "Is Bimetric Gravity Really Ghost Free?" (2013), arXiv:1301.3296.

[38] V. O. Soloviev, M. V. Tchichikina, "Bigravity in Kuchar's Hamiltonian formalism. 1. The General Case" (2012), arXiv:1211.6530.

[39] V. O. Soloviev, M. V. Tchichikina, "Bigravity in Kuchar's Hamiltonian formalism. 2. The special case" (2013), arXiv:1302.5096.

II

Condensed Matter, Particle and Radiation Physics

Entanglement and correlation energy in the lowest excited configuration of harmonium

Carlos L. Benavides-Riveros

Departamento de Física Teórica, Universidad de Zaragoza,
Pedro Cerbuna 12, 50009 Zaragoza, Spain

E-mail: carlobe@unizar.es

Abstract

Since the very early years of Quantum Chemistry, the harmonium has been regarded as an exactly solvable model for studying quantum-chemical properties. In this work we study in depth its first excited configuration, computing its entanglement measure and its correlation energy as well.

1. Introduction

One of the scientific dreams of Quantum Chemistry is to replace the wave function of electronic systems by the reduced 2-body density matrix γ_2, because it carries all necessary information required for calculating the quantum properties of atoms and molecules [1]. However, to date the N-representability problem for this matrix has proved to be a major challenge for theoretical quantum chemistry and in general it is not possible to characterize the set of admissible 2-body density matrices.

Two-electron systems are special cases in that γ_2 can be written "almost exactly" in terms of γ_1. Let us express γ_1 by means of the spectral theorem in terms of its set of natural orbitals $\{\phi_i\}$ as well as its set of occupation numbers $\{n_i\}$. For instance, the ground state of such a system admits a 1-density matrix:

$$\gamma_1(\boldsymbol{x}, \boldsymbol{x}') = (\uparrow_1\uparrow_{1'} + \downarrow_1\downarrow_{1'})\gamma_1(\boldsymbol{r}, \boldsymbol{r}') = (\uparrow_1\uparrow_{1'} + \downarrow_1\downarrow_{1'})\sum_i n_i\,\phi_i(\boldsymbol{r})\phi_i^*(\boldsymbol{r}'). \quad (1)$$

Here $\sum_i n_i = 1$, $r \in \mathbb{R}^3$ and we use the customary quantum-chemical notation $x :=$ (r, ς). The corresponding 2-density matrix is given by the expression:

$$\gamma_2(x_1, x_2; x_1', x_2') = (\uparrow_1\downarrow_2 - \downarrow_1\uparrow_2)(\uparrow_{1'}\downarrow_{2'} - \downarrow_{1'}\uparrow_{2'}) \sum_{ij} \frac{c_i c_j}{2} \phi_i(r_1)\phi_i(r_2)\phi_j^*(r_1')\phi_j^*(r_2')$$

$$\text{with coefficients} \quad c_i = \pm\sqrt{n_i}. \tag{2}$$

Although the expression is exact, the signs of the c_i need to be determined to find the ground state. Note that $\gamma_2^2 = \gamma_2$. The first excited state of the system admits a reduced 1-density matrix of the following kind:

$$\gamma_1(x; x') = \text{spin}_1 \times \sum_i n_i \left(\phi_{2i}(r)\phi_{2i}^*(r') + \phi_{2i+1}(r)\phi_{2i+1}^*(r')\right) \tag{3}$$

with $\sum_i n_i = 1$ and $\text{spin}_1 \in \{\uparrow_1\uparrow_{1'}, \frac{1}{2}(\uparrow_1\uparrow_{1'} + \downarrow_1\downarrow_{1'}), \downarrow_1\downarrow_{1'}\}$. The corresponding spinless 2-density matrix $\gamma_2(r_1, r_2; r_1', r_2')$ is given by

$$\sum_{ij} \frac{c_i c_j}{2} \left[\phi_{2i}(r_1)\phi_{2i+1}(r_2)\phi_{2j}^*(r_1')\phi_{2j+1}^*(r_2') + \phi_{2i+1}(r_1)\phi_{2i}(r_2)\phi_{2j+1}^*(r_1')\phi_{2j}^*(r_2')\right.$$

$$\left. - \phi_{2i}(r_1)\phi_{2i+1}(r_2)\phi_{2j+1}^*(r_1')\phi_{2j}^*(r_2') - \phi_{2i+1}(r_1)\phi_{2i}(r_2)\phi_{2j}^*(r_1')\phi_{2j+1}^*(r_2')\right],$$

$$\text{with coefficients} \quad c_i = +\sqrt{n_i}.$$

Due to the antisymmetry of this state, there is no ambiguity in the choice of the sign.

Aimed at studying the spectral problem of two-electron atoms, in the early years of the Quantum Theory, Heisenberg invented an exactly solvable model [2], here called *harmonium*. It exhibits two fermions interacting with an external harmonic potential and repelling each other by a Hooke-type force. Although it seems to be an over-simplificated problem, several questions related with it are analytically solvable, and wherefore it is tantalizing to use it in order to test the behavior of other useful and more realistic models. In fact, there is considerable recent work on studying and learning from harmonium, including correlation [3–5], approximation of functionals [6, 7], questions of entanglement and black hole entropy [8, 9]. Within the context of a phase-space density functional theory [10], here called WDFT, the alternating choice of signs in (2) has been shown to be the correct one for the harmonium's ground state [11, 12].

In this paper, we aim to study the first excited configuration of this system, emphasizing in its entanglement measure and correlation energy. The paper is divided as follows: in the second section we introduce the harmonium model, the third is devoted to the theory of Wigner natural orbitals in which this work is based, the fourth and fifth sections concern with the Schmidt decomposition of the triplet state of a system of two electrons. Finally, we conclude with an analysis of the entanglement and the correlation of such a system.

2. The atom studied

The general atom studied is a system of 2 interacting fermions in a harmonic well. Its Hamiltonian, in Hartree-like units, is [4, 13]:

$$H = \frac{p_1^2}{2} + \frac{p_2^2}{2} + \frac{k}{2}(r_1^2 + r_2^2) - \frac{\delta}{4}r_{12}^2, \qquad (4)$$

where $r_{ij} := |r_i - r_j|$. The interesting fact about this model is that the Hamiltonian can be reorganized in a completely separable one by means of introducing the so-called extracule and intracule coordinates, respectively given by

$$R = \frac{1}{\sqrt{2}}(r_1 + r_2), \qquad r = \frac{1}{\sqrt{2}}(r_1 - r_2),$$

$$P = \frac{1}{\sqrt{2}}(p_1 + p_2), \qquad p = \frac{1}{\sqrt{2}}(p_1 - p_2).$$

Thus, the Hamiltonian of such a system of two harmonically trapped fermions traduces to the composition of one extracular part and other intracular one:

$$H = H_R + H_r := \frac{P^2}{2} + \frac{\omega^2 R^2}{2} + \frac{p^2}{2} + \frac{\mu^2 r^2}{2}, \qquad (5)$$

where $\omega^2 = k$ and $\mu^2 = \omega^2 - \delta$.

3. Natural Wigner orbitals

Let us take any interference operator $|\Psi\rangle\langle\Phi|$ acting on the Hilbert space of a two-electron system; we denote

$$P_{2\Psi\Phi}(r_1, r_2; p_1, p_2; \varsigma_1, \varsigma_2; \varsigma_{1'}, \varsigma_{2'}) \qquad (6)$$

$$:= \int \Psi(r_1 - z_1, r_2 - z_2; \varsigma_1, \varsigma_2)\Phi^*(r_1 + z_1, r_2 + z_2; \varsigma_{1'}, \varsigma_{2'})e^{2i(p_1 \cdot z_1 + p_2 \cdot z_2)}\, dz_1\, dz_2.$$

In the particular case when $\Psi = \Phi$ we speak of Wigner quasiprobabilities, which are always real, and we write d_2 for P_2. The corresponding reduced 1-body functions are found by

$$P_{1\Psi\Phi}(r_1; p_1; \varsigma_1; \varsigma_{1'}) = 2 \int P_{2\Psi\Phi}(r_1, r_2; p_1, p_2; \varsigma_1, \varsigma_2; \varsigma_{1'}, \varsigma_2)\, dr_2\, dp_2\, d\varsigma_2. \qquad (7)$$

When $\Psi = \Phi$ we write d_1 for P_1. The associated spinless quantities $d_2(r_1, r_2; p_1, p_2)$ and $d_2(r; p)$ are obtained by tracing on the spin variables. The marginals of d_2 give the pair densities $\rho_2(r_1, r_2)$, $\pi_2(p_1, p_2)$. The marginals of d_1 give the electronic density, namely $\rho(r_1) = \int d_1(r_1, p_1) dp_1$, and the momentum density $\pi(p_1) = \int d_1(r_1, p_1) dr_1$. Putting together (2) and (1) with (6), one arrives [11] at:

$$d_1(r_1; p_1; \varsigma_1, \varsigma_{1'}) = 2 \int d_2(r_1, r_2; p_1, p_2; \varsigma_1, \varsigma_2; \varsigma_{1'}, \varsigma_2) d\varsigma_2 dr_2 dp_2$$

$$= (\uparrow_1 \uparrow_{1'} + \downarrow_1 \downarrow_{1'}) \sum_i n_i \chi_i(r_1; p_1).$$

Here n_i are the occupation numbers with $\sum_i n_i = 1$, the χ_{ij} the natural Wigner interferences and $\chi_i := \chi_{ii}$ denote the natural Wigner orbitals. It is clear that the factor $(\uparrow_1 \uparrow_{1'} + \downarrow_1 \downarrow_{1'})$ is a rotational scalar. The relation $c_i = \pm\sqrt{n_i}$ holds. In order to completely solve the N-representability problem remains the problem of determining the signs of the infinite set of square roots, to find the ground state.

4. The triplet state and its Schmidt decomposition

The spatial function for the ground state of harmonium is symmetric, and consequently its spin part is antisymmetric. For the first excited state the situation is exactly the opposite: the spatial function is antisymmetric and its spin part is symmetric. The triplet states are of the form [14]:

$$\Psi_{tk}(r_1, r_2; \varsigma_1, \varsigma_2) = (\text{spin}_2)_k \sum_{ij} \frac{1}{2} c_{ij} [\psi_i(r_1)\psi_j(r_2) - \psi_j(r_1)\psi_i(r_2)],$$

where $k \in \{1, 0, -1\}$, $c_{ij} = -c_{ji}$ and $(\text{spin}_2)_1 = \uparrow_1 \uparrow_2$, $(\text{spin}_2)_0 = \frac{1}{\sqrt{2}}(\uparrow_1 \downarrow_2 + \downarrow_1 \uparrow_2)$ and $(\text{spin}_2)_{-1} = \downarrow_1 \downarrow_2$. Here $\{\psi_i\}$ is a complete orthonormal set. In the absence of magnetic fields, the wave functions can be taken real. We thus assume that the matrix $C = [c_{ij}]$ is real, as well as the functions ψ_i. Wave function normalization gives rise to $\text{Tr}(C^T C) = \sum_{ij} c_{ij}^2 = 1$.

Hence, the spinless Wigner 2-body quasiprobability is given by the expression

$$d_2(r_1, r_2; p_1, p_2) = \frac{1}{4} \sum_{ij,kl} c_{ij} c_{kl} [P_{ik}(r_1; p_1)P_{jl}(r_2; p_2) - P_{il}(r_1; p_1)P_{jk}(r_2; p_2)$$
$$- P_{jk}(r_1; p_1)P_{il}(r_2; p_2) + P_{jl}(r_1; p_1)P_{ik}(r_2; p_2)], \qquad (8)$$

and integrating out one set of coordinates, we obtain the 1-body quasiprobability:

$$d_1(r; p) = 2 \int d_2(r, r_2; p, p_2) dr_2 dp_2 = 2 \sum_{ij} d_{ij} P_{ij}(r; p),$$

where $D = CC^{\mathrm{T}} = -C^2$ is a positive definite matrix. Since C is a real antisymmetric square matrix, there exists a real orthogonal matrix $Q = [q_{ij}]$ such that $A = Q^{\mathrm{T}}CQ$, with A a real block-diagonal matrix of the sort [15]:

$$A = \mathrm{diag}[A_0, A_1, \dots], \qquad A_0 = 0, \qquad A_i = \begin{pmatrix} 0 & a_i \\ -a_i & 0 \end{pmatrix}. \qquad (9)$$

By convention, here $a_i \geq 0$. Therefore

$$\sum_{ij,kl} c_{ij} c_{kl} P_{ik}(r_1; p_1) P_{jl}(r_2; p_2) = \sum_{ij,kl,vw} a_v a_w \left[q_{i,2v} q_{j,2v+1} - q_{i,2v+1} q_{j,2v} \right]$$

$$\times \left[q_{k,2w} q_{l,2w+1} - q_{k,2w+1} q_{l,2w} \right] P_{ik}(r_1; p_1) P_{jl}(r_2; p_2).$$

Let us now make the definition

$$\chi_{rp}(r; p) := \sum_{mk} q_{mr} P_{mk}(r; p) q_{kp}, \quad \text{so that} \quad P_{mk}(r; p) = \sum_{rp} q_{mr} \chi_{rp}(r; p) q_{kp}. \qquad (10)$$

This is the set of Wigner natural orbitals, and has the following nice property:

$$\int \chi_{rp}(r; p) \, dr \, dp = \delta_p^r. \qquad (11)$$

Hence,

$$\sum_{ij,kl} c_{ij} c_{kl} P_{ik}(r_1; p_1) P_{jl}(r_2; p_2)$$

$$= \sum_{vw} a_v a_w \left[\chi_{2v,2w}(r_1; p_1) \chi_{2v+1,2w+1}(r_2; p_2) - \chi_{2v,2w+1}(r_1; p_1) \chi_{2v+1,2w}(r_2; p_2) \right.$$

$$\left. - \chi_{2v+1,2w}(r_1; p_1) \chi_{2v,2w+1}(r_2; p_2) + \chi_{2v+1,2w+1}(r_1; p_1) \chi_{2v,2w}(r_2; p_2) \right].$$

The other three summands in (8) yield the same expression as the first summand. Then use symmetry under the interchange of the two particles. In summary,

$$d_2(r_1, r_2; p_1, p_2)$$

$$= \sum_{vw} a_v a_w \left[\chi_{2v,2w}(r_1; p_1) \chi_{2v+1,2w+1}(r_2; p_2) - \chi_{2v,2w+1}(r_1; p_1) \chi_{2v+1,2w}(r_2; p_2) \right.$$

$$\left. - \chi_{2v+1,2w}(r_1; p_1) \chi_{2v,2w+1}(r_2; p_2) + \chi_{2v+1,2w+1}(r_1; p_1) \chi_{2v,2w}(r_2; p_2) \right].$$

The reduced 1-body phase space (spinless) quasidensity for the triplet is obtained, as before,

$$d_1(r; p) = 2 \int d_2(r, r_2; p, p_2) \, dr_2 \, dp_2 = 2 \sum_w a_w^2 \left[\chi_{(2w,2w)}(r; p) + \chi_{(2w+1,2w+1)}(r; p) \right].$$

Notice that here each occupation number $n_i := 2a_i^2$ appears twice. This is a consequence of the Pauli exclusion principle. Unlike the singlet case, there is no sign rule to be deciphered here. Instead one has some ambiguities in the definition of the natural orbitals, noticed already by Löwdin and Shull [14].

5. Spectral analysis

The spectrum of the energy for harmonium is obviously $\left(\mathbb{N} + \frac{3}{2}\right)\omega + \left(\mathbb{N} + \frac{3}{2}\right)\mu$. Let us suppose that $\mu < \omega$, so that the energy of the first excited state is $E_{fs} = (3\omega + 5\mu)/2$. For our present purposes, it is enough to choose an intracule excitation state along the x-axis (say). The 2-quasidensity is given by:

$$\frac{2}{\pi^6} \exp\left(-\frac{2H_R}{\omega}\right) \exp\left(-\frac{2H_r}{\mu}\right) \left(\frac{(p_{1x} - p_{2x})^2 + \mu^2(x_1^2 - x_2^2)^2}{\mu} - \frac{1}{2}\right). \tag{12}$$

Henceforth from now we work in the non-trivial dimension, since the problem factorizes completely. By integrating one set of variables, the reduced 1-body spinless quasidensity is obtained:

$$d_1(r;p) = 2 \int d_2(r, r_2; p, p_2)\, dr_2\, dp_2 = \frac{2}{\pi} \left(\frac{2\sqrt{\omega\mu}}{\omega + \mu}\right)^3 e^{-\frac{2\omega\mu}{\omega+\mu}r^2 - \frac{2}{\omega+\mu}p^2} \left(\omega r^2 + \frac{1}{\omega}p^2\right), \tag{13}$$

after some work. In order to obtain the occupation numbers of this system, first we have to change the coordinates. Let us perform the transformation

$$(Q, P) := \left((\omega\mu)^{1/4}r, (\omega\mu)^{-1/4}p\right); \quad \text{or, in shorthand,} \quad U = Su, \tag{14}$$

where S is symplectic and $u = (r, p)$. We may also write $\vartheta := \arctan(P/Q)$, so that

$$P = U \sin\vartheta \quad \text{and} \quad Q = U \cos\vartheta. \tag{15}$$

Defining $t = (\sqrt{\omega} - \sqrt{\mu})/(\sqrt{\omega} + \sqrt{\mu})$, the 1-quasidensity (13) takes the simple form:

$$d_1(U, \vartheta) := d_1(u(U, \vartheta)) = \frac{2(1 - t^2)^3}{\pi(1 + t^2)^3} e^{-(1-t^2)U^2/(1+t^2)}U^2 \left(\frac{1 + t^2}{1 - t^2} + \frac{2t}{1 - t^2}\cos 2\vartheta\right),$$

and this can be expanded as follows:

$$d_1(U, \vartheta) = \sum_{rs} f_{rs}(U, \vartheta)d_{rs} \quad \text{where} \quad d_{rs} = 2\pi \int d_1(U, \vartheta)f_{rs}^*(U, \vartheta)U\, dU\, d\vartheta. \tag{16}$$

The functions f_i determine up to a phase the interferences: for $j \geq k$,

$$f_{jk}(x, p_x) = \frac{1}{\pi} (-1)^k \frac{\sqrt{k!}}{\sqrt{j!}} \left(2\sqrt{\omega\mu}\, x^2 + 2p_x^2/\sqrt{\omega\mu} \right)^{(j-k)/2}$$

$$\times e^{-i(j-k)\vartheta} L_k^{j-k} \left(2\sqrt{\omega\mu}\, x^2 + 2p_x^2/\sqrt{\omega\mu} \right) e^{-\sqrt{\omega\mu}\, x^2 - p_x^2/\sqrt{\omega\mu}},$$

where $\vartheta := \arctan\left(p_x/\sqrt{\omega\mu}\, x\right)$. The L_k^{j-k} are associated Laguerre polynomials. The f_{kj} are complex conjugates of the f_{jk}. Then, for $r \geq s$,

$$2\pi \int f_{rs}^*(U, \vartheta) \, d_1(U, \vartheta) \, U \, dU \, d\vartheta$$

$$= \frac{4(1-t^2)^3}{\pi(1+t^2)^3} (-1)^s \frac{\sqrt{s!}}{\sqrt{r!}} \int_0^\infty e^{-(1-t^2)U^2/(1+t^2)} e^{-U^2} (2U^2)^{(r-s)/2} L_s^{r-s}(2U^2) U^3 \, dU$$

$$\times \pi \left[\frac{2(1+t^2)}{1-t^2} \delta_r^s + \frac{2t}{1-t^2} (\delta_r^{s+2} + \delta_r^{s-2}) \right],$$

so that

$$d_1(U, \vartheta) = \sum_s d_{ss}(t) f_{ss}(U, \vartheta) + d_{s+2,s}(t) f_{s+2,s}(U, \vartheta) + d_{s,s+2}(t) f_{s,s+2}(U, \vartheta), \quad (17)$$

where actually $d_{s+2,s} = d_{s,s+2}$. Using the standard Mellin transform [16], we obtain by fairly easy manipulations,

$$d_{ss}(t) = (1-t^2)^2 \left(s\, t^{2s-2} + (1+s)\, t^{2s} \right);$$

$$d_{s,s+2}(t) = (1-t^2)^2 \sqrt{(s+1)(s+2)}\, t^{2s+1}.$$

This means that, to find the occupation numbers, one has to diagonalize a symmetric pentadiagonal matrix:

$$D = (1-t^2)^2 \begin{pmatrix} 1 & 0 & \alpha_0 t & 0 & \cdots \\ 0 & 1+2t^2 & 0 & \alpha_1 t^3 & \cdots \\ \alpha_0 t & 0 & 2t^2 + 3t^4 & 0 & \cdots \\ 0 & \alpha_1 t^3 & 0 & 3t^4 + 4t^6 & \cdots \\ \vdots & \vdots & \vdots & \vdots & \ddots \end{pmatrix}, \quad (18)$$

where $\alpha_s := \sqrt{(s+1)(s+2)}$. It is easy to check that the eigenspaces split into two parts: $\ell_2 = V_1 \oplus V_2$, where $V_1 = \{x : \text{all } x_{2n} = 0\}$ and $V_2 = \{x : \text{all } x_{2n+1} = 0\}$. It is worth emphasizing that these matrices have the same set of eigenvalues, which is a consequence of the fact that the occupation numbers must appear twice, as we already showed in the last section 4.

Figure 1 shows the behavior of the rank-eight approximation of the eigenvalues, as the parameter t is varied. The first eigenvalue is very close to 1 in the neighborhood of $t = 0$, while the others are very small. As the value of t rises, the first eigenvalue begins to decrease and the others rise for a while. In the neighborhood of $t = 1$ all eigenvalues approach zero.

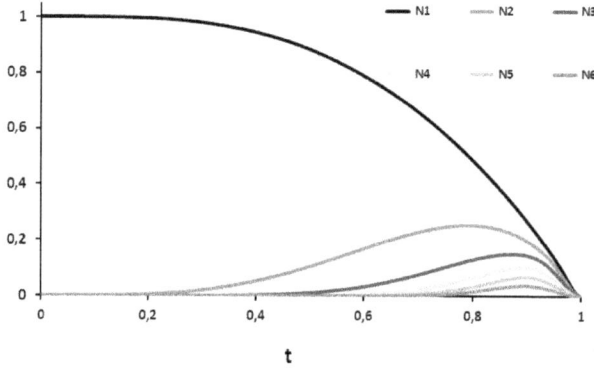

Figure 1.: First six eigenvalues of the matrix D_{even} [16].

It is worth noticing that t is a very nonlinear parameter: although $t \sim \delta/8k$ for small δ, the value $t = 1/2$ means $\mu/\omega = 1/9$ or $\delta/k = 80/81$. This illustrates the fact that, unless δ is pretty close to the dissociation value $t = 1$, the excited harmonium is well described by the Hartree-Fock state. Moreover, whenever $t \lesssim 0.6$, that is, $\delta/k \lesssim 255/256$, the first two occupation numbers contain almost all the physical information for the system.

6. Entanglement and correlation

We are in position to compare the triplet system with the singlet one in regard to entanglement and correlation energy. A useful quantity is the linear entropy s associated to the 1-body function:

$$s = 1 - \Pi(d_1),\tag{19}$$

where $\Pi(d_1)$ is the *purity* of the system – see below. Mathematically, the quantity s is a lower bound for the Jaynes entropy, which has been used to quantify the entanglement between one particle and the other $N - 1$ particles of the system [17], and proposed as a handle on the correlation energies [18]. In this paper the singlet has been modeled in such a way that, for each one-dimensional mode:

$$\Pi_{\text{gs},1}(d_1) = \int d_1^2(r;p)\,\mathrm{d}r\,\mathrm{d}p = \sum_i n_i^2.\tag{20}$$

Instead, for the triplet one should take for the excited mode:

$$\Pi_{\text{fs},x}(d_1) = \frac{1}{2} \int d_1^2(r_x; p_x)\,dr_x\,dp_x = \sum_i n_i^2. \tag{21}$$

This second definition is natural in that correlations due solely to the antisymmetric character of the wave function *do not contribute* to the entanglement of the system. This ensures that the entropy for a 1-body function of the Hartree-Fock type is zero.

In the singlet case, the occupation numbers are equal to $(1 - t^2)\,t^{2i}$ (see [12]). Thus, the purity of this system is easily computable, to wit, $\Pi_{\text{gs},1}(d_1) = (1 - t^2)/(1 + t^2)$ for each mode. Note that this quantity coincides with the quotient of the geometric and arithmetic means of the frequencies, that is, $\Pi_{\text{gs},1} = 2\sqrt{\omega\mu}/(\omega + \mu)$. For the triplet state, we get

$$\text{Tr}(d_1^2) = 2(1 - t^2)^4 \sum_{i=1}^{\infty} \left[2i(i+1)\,t^{2(2i-1)} + i^2\,t^{4(i-1)} \right] = \frac{2(1 - t^2)}{1 + t^2}\left[1 + \frac{2t^2}{(1 + t^2)^2} \right],$$

after some calculation. So the purity of the first excited mode is

$$\Pi_{\text{fs},x} = \frac{1 - t^2}{1 + t^2}\left[1 + \frac{2t^2}{(1 + t^2)^2} \right] = \Pi_{\text{gs},1}\left[1 + \frac{2t^2}{(1 + t^2)^2} \right] = \frac{2\sqrt{\omega\mu}}{\omega + \mu}\left(1 + \frac{1}{2}\left(\frac{\omega - \mu}{\omega + \mu}\right)^2 \right). \tag{22}$$

Since the other two modes contribute with two ground state factors, the total purity can be written as $\Pi_{\text{fs}} = \Pi_{\text{fs},x}\Pi_{\text{gs},y}\Pi_{\text{gs},z}$. For the purity parameter, one obtains finally

$$s_{\text{gs}} = 1 - \left(\frac{1 - t^2}{1 + t^2}\right)^3 \quad \text{and} \quad s_{\text{fs}} = 1 - \left(\frac{1 - t^2}{1 + t^2}\right)^3\left[1 + \frac{2t^2}{(1 + t^2)^2} \right] = s_{\text{gs}} - \frac{2t^2(1 - t^2)^3}{(1 + t^2)^5}. \tag{23}$$

In conclusion, $s_{\text{fs}} \leq s_{\text{gs}}$.

Finally, we compute, for the first time, the correlation energy of the *excited* harmonium. The Hartree-Fock approximation for the relevant mode, in view of (8), is of the form

$$W_{\text{HF}}(r_1, r_2; p_1, p_2) = \frac{1}{2}\left[W_{00}(r_1; p_1)W_{11}(r_2; p_2) - W_{01}(r_1; p_1)W_{10}(r_2; p_2) \right.$$
$$\left. - W_{10}(r_1; p_1)W_{01}(r_2; p_2) + W_{11}(r_1; p_1)W_{00}(r_2; p_2) \right],$$

where $W_{00}(r; p) = \frac{1}{\pi}e^{-\eta r^2 - p^2/\eta}$, $\qquad W_{11}(r; p) = \frac{2}{\pi}e^{-\eta r^2 - p^2/\eta}\left(\eta r^2 + p^2/\eta - \frac{1}{2} \right),$

with their corresponding interferences. Remember that $\int W_{ij}\,dr\,dp = \delta_{ij}$. In intracule-extracule coordinates:

$$W_{\text{HF}}(R, r; P, p) = \frac{2}{\pi^2}\left(\eta r^2 + p^2/\eta - \frac{1}{2} \right) e^{-\eta R^2 - P^2/\eta - \eta r^2 - p^2/\eta}. \tag{24}$$

The parameter η is determined by minimization. The mean value of the energy predicted by this function is:

$$E_{\mathrm{HF}} = \frac{1}{2} \int (p^2 + \omega^2 r^2)[W_{00}(r;p) + W_{11}(r;p)] \, dr \, dp - \frac{\delta}{4} \int (r_1 - r_2)^2 \, W_{\mathrm{HF}}(1,2) \, d1 \, d2$$

$$= \left(\eta + \frac{\omega^2}{\eta} \right) - \frac{3\delta}{4\eta} = \eta + \frac{\omega^2 + 3\mu^2}{4\eta}.$$

The minimum $dE/d\eta = 0$ occurs when $\eta = \frac{1}{2}\sqrt{\omega^2 + 3\mu^2}$. Therefore, the energy predicted by Hartree-Fock is $\sqrt{\omega^2 + 3\mu^2}$. Thus, the "correlation energy" for the lowest excited state of harmonium is:

$$E_{\mathrm{c,fs}} = E_{\mathrm{fs}} - E_{\mathrm{HF}} = \frac{3\omega + 5\mu}{2} - \sqrt{\omega^2 + 3\mu^2} - 2\sqrt{(\omega^2 + \mu^2)/2} \sim -\frac{7}{64} \frac{\delta^2}{\omega^3}. \quad (25)$$

Thus, the relative correlation energies are

$$\mathcal{E}_{\mathrm{fs}} := \frac{|E_{\mathrm{c,fs}}|}{E_{\mathrm{fs}}} \sim \frac{7}{256} \frac{\delta^2}{\omega^4} \quad \text{and} \quad \mathcal{E}_{\mathrm{gs}} := \frac{|E_{\mathrm{c,gs}}|}{E_{\mathrm{gs}}} \sim \frac{1}{32} \frac{\delta^2}{\omega^4}. \quad (26)$$

Both quantities are related by a factor of $7/8$. For this approximation, as one would have expected, $\mathcal{E}_{\mathrm{fs}} \leq \mathcal{E}_{\mathrm{gs}}$.

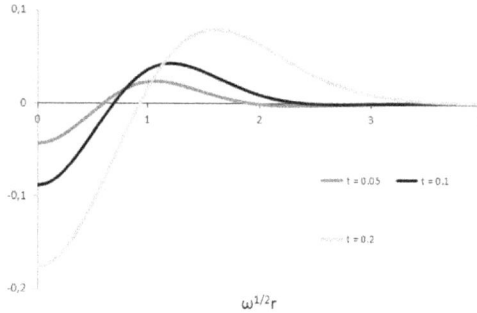

Figure 2.: Harmonium's hole for the triplet: $(\rho(r) - \rho_{\mathrm{HF}}(r))/\omega^{1/2}$ as a function of $\omega^{1/2}r$ [16].

Finally, Figure 2 shows the difference between the exact profile 1-density and the Hartree-Fock profile 1-density for the harmonium triplet,

$$\rho_{\mathrm{HF}}(r) := \int W_{\mathrm{HF}}(r, r_2; p_1, p_2) \, dp_1 \, dr_2 \, dp_2. \quad (27)$$

The "Harmonium's hole" observed in the neighborhood of $r = 0$ graphically shows the Hartree-Fock underestimation of the mean distance between the fermions, for the excited configuration of harmonium as well.

Acknowledgements

I thank the organizers of the Barcelona Postgrad Encounters on Fundamental Physics for their warm hospitality in Barcelona. This work has been supported by a "Francisco José de Caldas" scholarship.

References

[1] C. A. Coulson, "Present state of molecular structure calculations", *Rev. Mod. Phys.* **32** (1960), 170.

[2] W. Heisenberg, "Mehrkörperproblem und Resonanz in der Quantenmechanik", *Z. Physik* **38** (1926), 411.

[3] B. P. Van Zyl, "Wigner distribution function for a harmonically trapped gas of ideal fermions and bosons at arbitrary temperature and dimensionality", *J. Phys. A: Math. Theor.* **45** (2012), 315302.

[4] M. Moshinsky, "How Good is the Hartree-Fock Approximation", *Am. J. Phys.* **36** (1968), 52.

[5] N. H. March et al., "Proposed definitions of the correlation energy density from a Hartree-Fock starting point: The two-electron Moshinsky model atom as an exactly solvable model", *Phys. Rev.* **A77** (2008), 042504.

[6] I. Nagy, J. Pipek, "Approximations for the interparticle interaction energy in an exactly solvable two-electron model atom", *Phys. Rev.* **A81** (2010), 014501.

[7] C. L. Benavides-Riveros, J. Várilly, "Testing one-body density functionals on a solvable model", *Eur.Phys. J.* **D66** (2012), 274.

[8] C. Schilling, D. Gross, M. Christandl, "Pinning of Fermionic Occupation Numbers", *Phys. Rev. Lett.* **110** (2013), 040404.

[9] M. Srednicki, "Entropy and area", *Phys. Rev. Lett.* **71** (1993), 666.

[10] J. P. Dahl, "Moshinsky atom and density functional theory — A phase space view", *Canadian J. Chem.* **87** (2009), 784.

[11] P. Blanchard, J. M. Gracia-Bondía, J. C. Várilly, "Density functional theory on phase space", *Int. J. Quant. Chem.* **112** (2012), 1134.

[12] K. Ebrahimi-Fard, J. M. Gracia-Bondía, "Harmonium as a laboratory for mathematical chemistry", English, *J. Math. Chem.* **50** (2012), 440.

[13] A. Calles, M. Moshinsky, "How Good Is the Hartree-Fock Approximation? II. The Case of Closed Shells", *Am. J. Phys.* **38** (1970), 456.

[14] P.-O. Löwdin, H. Shull, "Natural Orbitals in the Quantum Theory of Two-Electron Systems", *Phys. Rev.* **101** (1956), 1730.

[15] R. A. Horn, C. R. Johnson, "Matrix Analysis", Cambridge University Press (Cambridge), 1985.

[16] C. L. Benavides-Riveros, J. M. Gracia-Bondía, J. C. Várilly, "Lowest excited configuration of harmonium", *Phys. Rev.* **A86** (2012), 022525.

[17] N. Helbig, I. V. Tokatly, A. Rubio, "Physical meaning of the natural orbitals: Analysis of exactly solvable models", *Phys. Rev.* **A81** (2010), 022504.

[18] G. T. Smith, H. L. Schmider, V. H. Smith, "Electron correlation and the eigenvalues of the one-matrix", *Phys. Rev.* **A65** (2002), 032508.

Nonlinear field theory with topological solitons: Skyrme models

C. Adam[1,a], C. Naya[2,a], J. Sanchez-Guillen[3,a] and A. Wereszczynski[4,b]

[a]Departamento de Física de Partículas, Universidade de Santiago de Compostela, and
Instituto Galego de Física de Altas Enerxías (IGFAE),
Facultade de Física-Campus Vida, 15782 Santiago de Compostela, Spain

[b]Instytut Fizyki, Uniwersytetu Jagiellońskiego, ul. Reymonta 4, Kraków, Poland

E-mail: [1]adam@fpaxp1.usc.es, [2]carlos.naya87@gmail.com, [3]joaquin@fpaxp1.usc.es,
[4]wereszczynski@th.if.uj.edu.pl

Abstract

In this talk, we give new insight into one of the best-known nonlinear field theories, the Skyrme model. We present some exact relevant solutions coming from different new versions (gauged BPS baby as well as vector BPS Skyrme models) giving rise to topological solitons, and highlighting the BPS character of the theory.

1. Introduction

In 1834, the Scottish naval engineer J. Scott Russell saw in the Edinburgh canal a solitary wave for the first time. Nowadays, they are known as solitons, and among their main features we can point out that they are localized solutions which conserve their form, even after collisions. One specific kind of these objects are the interesting topological solitons. They are given by the collective excitations of the relevant degrees of freedom of the system and their existence and stability are caused by the global topological structure of the base and field space.

It is here where the Skyrme model appears as a nonlinear field theory, introduced to describe nuclei, supporting topological solitons [1–3]. The main feature of this model is that it considers the primary fields as pions whereas nucleons and nuclei are described by collective nonlinear excitations of the degrees of freedom, i.e., by topological solitons. One of the most clever ideas of Skyrme was to identify the topological charge which arises in the model with the baryon number ensuring the latter to be conserved. Although this is a theory in 3+1 dimensions, due to its nonlinear character and complexity, sometimes it is important to study its *little brother* in 2+1 too, the Baby Skyrme model [4, 5], since it is simpler and can give us some clues to deal with the original model. This is an advice that we will follow in the present talk where two of the three different models are 2+1 dimensional theories.

The Standard Skyrme Model consists of two terms: a quadratic term in first derivatives (the nonlinear sigma-model term), and a quartic one (the so-called Skyrme term). Both of them are necessary for the stability of the solutions from the point of view of the Derrick scaling, however, we can generalize the model with two more terms asking to have terms at most quadratic in time derivatives (so we have a standard Hamiltonian formulation). Thus, we can add a potential and a sextic term, so the more general model is given by

$$\mathscr{L}_{\mathrm{Skr}} = \mathscr{L}_2 + \mathscr{L}_4 + \mathscr{L}_6 + \mathscr{L}_0, \tag{1}$$

with

$$\mathscr{L}_2 = -\frac{f_\pi^2}{4}\,\mathrm{tr}(L_\mu L^\mu), \qquad \mathscr{L}_4 = -\frac{1}{32e^2}\,\mathrm{tr}\left([L_\mu, L_\nu]^2\right),$$
$$\mathscr{L}_6 = \lambda^2 \pi^4 B_\mu^2, \qquad \mathscr{L}_0 = -\mu^2 V,$$

and

$$L_\mu = U^\dagger \partial_\mu U, \qquad B^\mu = \frac{1}{24\pi^2}\,\mathrm{tr}(\epsilon^{\mu\nu\rho\sigma} U^\dagger \partial_\nu U U^\dagger \partial_\rho U U^\dagger \partial_\sigma U),$$

where B^μ is the topological current and the degrees of freedom are the U fields ($U \in$ SU(2)). This effective field theory is related to QCD in the large N (number of colours) limit, which does not favour any particular term. Another important comment on this Skyrme model is that although originally it was thought for Nuclear and Particle Physics, now it has applications in other branches, as for instance the planar Skyrmions in ferromagnetic materials: a collective excitation of spins from the homogeneous case.

Just to conclude this brief introduction we will present a key concept in our work, the BPS Bound, named after Bogomolńyi, Prasad and Sommerfield. It consists in a lower energy bound of the system depending on its topology and not on the field configuration. Solutions saturating the bound (BPS solutions) fulfill a set of first order equations called BPS equations. Thus, BPS solutions imply a simplification of the field equations going from second to first order. Furthermore, when they saturate the bound, they minimize the energy, so BPS states are stable. An important characteristic for the Skyrme model is that the bound is related to topological quantities as the winding number: $E \propto n$. Unfortunately, the Standard Skyrme model cannot have solutions saturating the bound. One case where this is possible is the BPS Skyrme Model [6] where we just take into account the sextic as well as the potential term:

$$\mathscr{L}_{BPS} = \mathscr{L}_6 + \mathscr{L}_0. \tag{2}$$

This introduces the idea of infinite massive (quenched) pions since we forget about the kinetic term $\mathscr{L}_2$. This restricted Skyrme model solves old problems of hadrons and nuclei, and is being actively used in their phenomenology [7–9]

With all these concepts in mind we are going to introduce three different new versions of the Skyrme model. In section 2, we will present a BPS baby model after introducing the gauge group U(1) with the usual Maxwell term for the gauge field. On the other hand, in section 3 we study the Skyrme field coupled to the vector meson ω, firstly in 2+1 dimensions to finally reach the 3+1 dimensional case.

2. Gauged BPS baby Skyrme model

The first model we are going to deal with is the gauged version of the BPS Skyrme Model above [6] but in 2+1 dimensions, the Gauged BPS baby Skyrme model [10]. The Lagrangian density is given by

$$\mathscr{L} = -\frac{\lambda^2}{4}(D_\alpha\boldsymbol{\phi} \times D_\beta\boldsymbol{\phi})^2 - \mu^2 V(\boldsymbol{n}\cdot\boldsymbol{\phi}) - \frac{1}{4g^2}F_{\alpha\beta}^2. \tag{3}$$

The degrees of freedom of the system are described by a vector of scalar fields $\boldsymbol{\phi} = (\phi_1, \phi_2, \phi_3)$ with unit length $\boldsymbol{\phi}^2 = 1$ and where the first term is quartic in first derivatives and at the same time is the square of the topological current, so it plays the rôle of the $\mathscr{L}_4$ and $\mathscr{L}_6$ in the 3+1 version. Moreover, the $\mathscr{L}_2$ term is not necessary for stability. On the other hand, since we have included the coupling to the gauge field, as well as the typical Maxwell term the usual derivatives have been replaced by the covariant ones:

$$D_\alpha\boldsymbol{\phi} = \partial_\alpha\boldsymbol{\phi} + A_\alpha\boldsymbol{n} \times \boldsymbol{\phi}. \tag{4}$$

Notice that if we ask for finite energy field configurations in the full baby model (the $\mathscr{L}_2$ term included), $\boldsymbol{\phi}(x, t)$ has to approach its vacuum value independently of the direction so the base space $\mathbb{R}^2$ can be compactified to a two-sphere S^2, and the field $\boldsymbol{\phi}$ is a map $S^2 \to S^2$ characterized by an integer winding number or topological degree

$$\deg[\boldsymbol{\phi}] = \frac{1}{4\pi}\int d^2x\, \boldsymbol{\phi} \cdot \partial_1\boldsymbol{\phi} \times \partial_2\boldsymbol{\phi} = k, \qquad k \in \mathbb{Z}. \tag{5}$$

Although this is not necessary for the BPS baby model we will assume it too because of consistency.

The next step, in order to simplify things, will be to choose $\boldsymbol{n} = (0, 0, 1)$ and the standard static ansatz:

$$\boldsymbol{\phi}(r, \varphi) = \begin{pmatrix} \sin f(r)\cos n\varphi \\ \sin f(r)\sin n\varphi \\ \cos f(r) \end{pmatrix} \tag{6}$$

$$A_0 = A_r = 0, \qquad A_\varphi = na(r), \tag{7}$$

so the electric and magnetic fields are

$$E_i = 0, \qquad B = \partial_1 A_2 - \partial_2 A_1 = \frac{na'(r)}{r}, \tag{8}$$

whereas the winding number is just $\deg[\phi] = n$. It will be also useful to introduce the new variable and field

$$y = r^2/2, \qquad h = \frac{1}{2}(1 - \cos f). \tag{9}$$

In this model, we will work with the so called old potential: $V_0 = 1 - \phi_3 = 2h$, a 2+1 version of the standard Skyrme potential used in 3+1 dimensions. Then, the boundary conditions are

$$h(0) = 1 \Leftrightarrow f(0) = \pi, \qquad a(0) = 0 \tag{10}$$

at the origin, and if we have compacton solutions

$$a_y(y = y_0) = 0, \qquad h(y_0) = h_y(y_0) = 0 \tag{11}$$

at the radius, or

$$\lim_{y \to \infty} h(y) = 0, \qquad \lim_{y \to \infty} a_y(y) = 0 \tag{12}$$

for exponentially or powerlike localized solutions. It can be shown that the old baby potential corresponds to compacton solutions. Although it seems we have five conditions notice that we are including a new parameter, the compacton radius y_0.

Before going to the solutions, we will ask ourselves if a BPS bound can be found for this model. Fortunately, the answer is positive. Thus, regarding the static energy functional and after a non-trivial manipulation we arrive at

$$E \geq E_0 \lambda^2 \int d^2x q W' = 4\pi |n| E_0 \lambda^2 \langle W' \rangle_{S^2} = 2\pi |n| E_0 \lambda^2 |W(\phi_3 = -1)|, \tag{13}$$

where we will call W the superpotential and n the winding number (making clear it is a topological bound). This bound will be saturated if the fields obey the BPS equations given by

$$Q = W', \tag{14}$$

$$B = -g^2 \lambda^2 W, \tag{15}$$

where

$$Q \equiv \phi \cdot D_1 \phi \times D_2 \phi = q + \epsilon_{ij} A_i \partial_j (n \cdot \phi). \tag{16}$$

Let us comment on the superpotential W present in this BPS bound. It appears in the process of deriving the bound that W is not a free function but is restricted to globally existing (on the unit interval) solutions of the following superpotential equation

$$\lambda^2 (W')^2 + g^2 \lambda^4 W^2 = 2\mu^2 V. \tag{17}$$

(a) Function h with its derivative (dashed line).

(b) Function a and its derivative (dashed line).

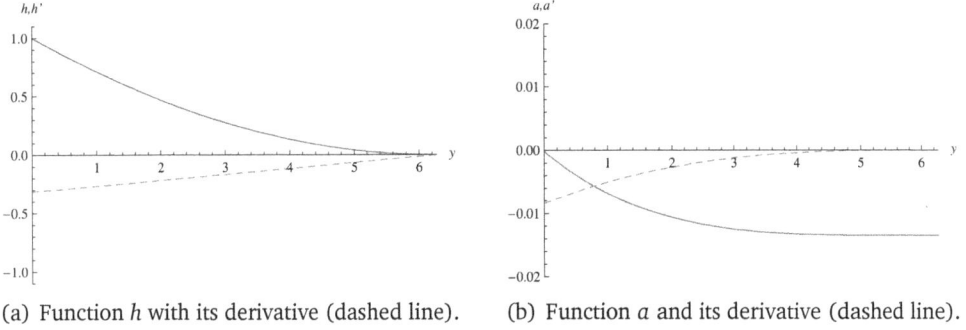

Figure 1.: Solutions for $g = 0.1$ and $\mu^2 = 0.1$.

We give it this name since it is very similar to the superpotential equation of fake supergravity [11]. In fact, we can find this kind of equation in self-gravitating domain walls, where the terms have different signs, or in extremal supersymmetric black holes, where they enter with the same sign. It is also necessary to remind that in order to have BPS solitons the solution of this superpotential equation has to exist globally. Since it depends on the potential V, we will not have a solution for all V. For instance, if V has two vacua there are no solitons. It is also important to remark that in this model all solutions of the static equations are BPS (we will see it with the numerical solutions).

In figures 1 and 2 we can see the numerical solutions we have found for different values of the parameter g (fixing $\lambda = n = 1$). In the case of the value of μ^2 we just show cases corresponding to $\mu^2 = 0.1$ since the situation is quite similar for other values. However, this does not happen when considering g. In figure 1 we see the solution with $g = 0.1$ (similar situation has been found for g lower than this.), where the derivative of the field a corresponding to the magnetic field is here localized at the center. On the other hand, looking now at figure 2 we see a different situation, with the magnetic field being almost constant until the compacton boundary and the field h approaching the zero value suddenly. In fact, the higher the value of g, the more pronounced is the approach to zero. Finally, using the numerical solutions for several values of g we can see that indeed all soliton solutions are BPS as shown in figure 3, where the solid line represents the calculated bound $E_B = 2\pi n\lambda^2 |W(h = 1)|$.

3. Vector Skyrme models

In the model above we have just partially answered the question of how solitons interact with the lightest fields. We say partially because besides the U(1) Maxwell field we can consider another one to couple to solitons: the omega vector mesons, which

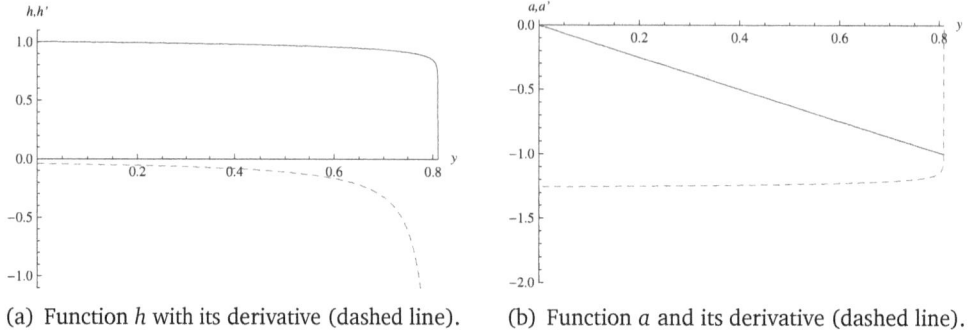

(a) Function h with its derivative (dashed line). (b) Function a and its derivative (dashed line).

Figure 2.: Solutions for $g = 2$ and $\mu^2 = 0.1$.

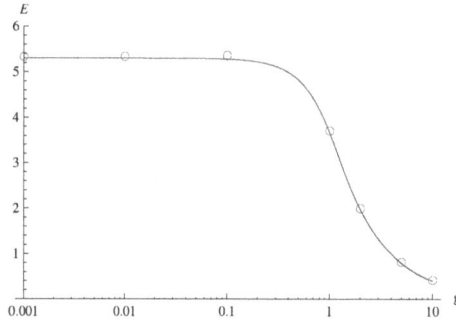

Figure 3.: Energy of static solutions (open circles) compared to the BPS bound (continuous line).

after pions, are the lightest ones and their (phenomenologically motivated) coupling is just $\omega_\mu B^\mu$. As we have commented before, the 2+1 version of the system can be a good starting point to then move on to the 3+1 dimensional case, and this will be the process we are going to follow here. Finally, since we are going to compare this vector model with its BPS brother without additional mesons, we will have to be aware of possible changes in the typical mass hierarchy: the BPS model is based on pions with infinite mass while here we are regarding that heavier mesons have finite mass.

3.1. Vector BPS baby Skyrme model

As it was commented above, the difference of this model with respect to the first one is just that now we couple the solitons to a vector meson instead of a gauge field. Then, the 2+1 dimensional version of the Skyrme term is replaced here by the vector meson coupling [12]:

$$\mathscr{L} = -\mu^2 V(\phi_3) - \frac{1}{4}(\partial_\mu \omega_\nu - \partial_\nu \omega_\mu)^2 + \frac{1}{2}M^2 \omega_\mu^2 + \lambda' \omega_\mu B^\mu, \tag{18}$$

where B^μ is the topological current given by

$$B^\mu = -\frac{1}{8\pi} \epsilon^{\mu\alpha\beta} \boldsymbol{\phi} \cdot (\partial_\alpha \boldsymbol{\phi} \times \partial_\beta \boldsymbol{\phi}), \tag{19}$$

and we will use the generalized old baby potentials

$$V = \left(\frac{1-\phi_3}{2}\right)^\alpha, \qquad \alpha \geq 1. \tag{20}$$

As before, we can see the vector field $\boldsymbol{\phi}$ as a map from S^2 to S^2 after compactification of the base space. Thus, we can use the stereographic projection

$$\boldsymbol{\phi} = \frac{1}{1+|u|^2}(u+\bar{u}, -\mathrm{i}(u-\bar{u}), |u|^2 - 1), \tag{21}$$

with the axially symmetric ansatz:

$$u = f(r)\mathrm{e}^{\mathrm{i}n\phi} \qquad \omega \equiv \omega = \omega(r) \qquad \omega_i = 0, \tag{22}$$

where n is the winding number. The boundary conditions to have solitons will be

$$\begin{array}{cccc}
f(r=0) = \infty, & f(r=R) = 0, & f'(r=R) = 0, \\
\omega'(r=0) = 0, & \omega(r=R) = 0, & \omega'(r=R) = 0,
\end{array} \tag{23}$$

where R will be finite or infinite depending on the potential.

Studying now solutions for the model we will focus on the most interesting case, the potential with $\alpha = 2$. It is so interesting because we have an analytical solution as well as a BPS bound. Then, defining the new field $g = \frac{f^2}{1+f^2}$ and variable $x = r^2/2 = \frac{n\lambda}{\sqrt{2}\mu}y$ (where λ is the initial λ' up to a constant), we can solve the equation of motion for g

$$g(y) = \frac{1}{\mathrm{K}_1(2/\beta)} \frac{\mathrm{K}_1\left(\frac{2}{\beta}\sqrt{1+\beta y}\right)}{\sqrt{1+\beta y}}, \tag{24}$$

with K_1 the modified Bessel function of the second type and $\beta = (2\mu)/(n\lambda M)$, whereas the equation for the vector meson is just

$$\omega_y = -\sqrt{2}\frac{\mu}{M}g. \tag{25}$$

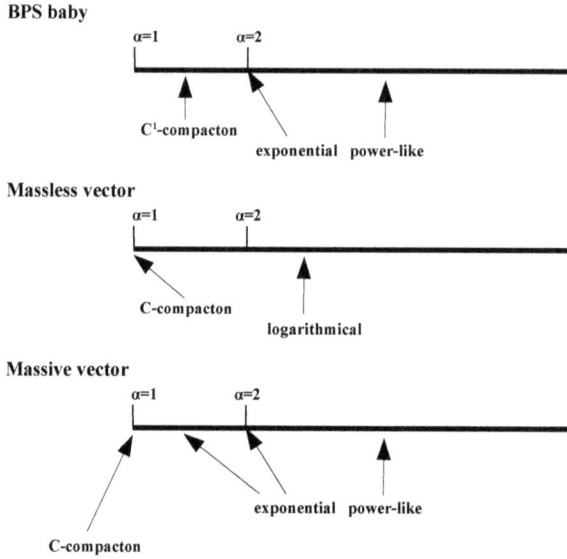

Figure 4.: Comparing solutions of the BPS baby Skyrme model with those of the baby vector one.

Finally, the BPS energy is

$$E = -\frac{\pi}{2}\left(\frac{n\lambda}{\mu}\right)^2 \omega(0)\omega_x(0) \simeq \sqrt{2}\pi\frac{n\lambda\mu}{M}\left(1 - \frac{1}{4}\beta + \cdots\right), \qquad (26)$$

which is nonlinear on n. Therefore, states with high n are unstable with respect to those with lower n.

Other results either for massive vector mesons or massless ones are compared with those for the BPS baby model in figure 4. The most important difference appears for $\alpha = 1$ since in this case, we get a C-compacton for the vector model. This means that at the boundary of the compacton there is a discontinuity, so that a Dirac delta is needed as a source term, what implies that the topological charge is screened. However, it is a positive difference since, because of the hierarchy problem, we have to remember that in the vector model the omega mesons are massless or have a finite mass whereas the lighter pions in nature are thought to be infinitely massive. Then, if the situation were the same the baby BPS Skyrme model would not be useful from the phenomenological point of view.

In the same way we could think about sending the mass of the omega mesons to infinity. With this we will recover the original mass hierarchy since if we assume that pions have infinite mass the same should stand for omegas. Then, our Lagrangian density (18) would become

$$\mathscr{L} = -\mu^2 V(\phi_3) + \frac{1}{2} M^2 \omega_\mu^2 + \lambda' \omega_\mu B^\mu \,, \tag{27}$$

so the field equation for ω mesons results in

$$\omega_\mu = -\frac{\lambda'}{M^2} B_\mu \,. \tag{28}$$

Plugging this back into the Lagrangian density we arrive at

$$\mathscr{L} = -\mu^2 V(\phi_3) - \frac{1}{2} \frac{(\lambda')^2}{M^2} B_\mu^2 \,. \tag{29}$$

Thus, in this limit of infinite massive vector mesons we actually get the BPS baby Skyrme model.

3.2. Vector BPS Skyrme model

The next and natural step is to use the knowledge we have from the baby model for the 3+1 dimensional version. Now, the corresponding Lagrangian density is [13]:

$$\mathscr{L} = -\mu^2 V(U, U^\dagger) - \frac{1}{4} (\partial_\mu \omega_\nu - \partial_\nu \omega_\mu)^2 + \frac{1}{2} M^2 \omega_\mu^2 + \lambda' \omega_\mu B^\mu \,, \tag{30}$$

where B^μ is the topological current given by

$$B^\mu = \frac{1}{24\pi^2} \operatorname{tr}(\epsilon^{\mu\nu\rho\sigma} U^\dagger \partial_\nu U U^\dagger \partial_\rho U U^\dagger \partial_\sigma U) \,, \tag{31}$$

and as before we will use the generalized Skyrme potentials, which in 3+1 dimensions are:

$$V = \left(\frac{1 - \operatorname{tr} U}{2} \right)^\alpha = (1 - \cos \xi)^\alpha \,. \tag{32}$$

Furthermore, similar to the situation we had in 2+1 dimensions, the corresponding fundamental field $U \in SU(2)$ defines now a map from the three-sphere in the compactified base space to the field space SU(2), a map which is characterized by a winding number. Then, a suitable parametrization for this U field is

$$U = e^{i\xi n \cdot \tau} = \cos \xi + i \sin \xi \, n \cdot \tau \,, \tag{33}$$

with

$$n = \frac{1}{1 + |u|^2} (u + \bar{u}, -i(u - \bar{u}), 1 - |u|^2) \,. \tag{34}$$

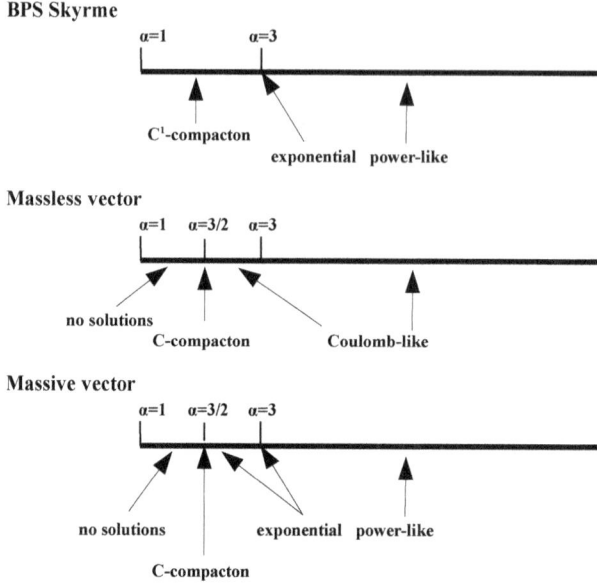

Figure 5.: Comparing solutions of the BPS Skyrme model with those of the vector one.

And as before, in order to simplify things, we will assume the static ansatz:

$$\omega_0 \equiv \omega = \omega(r), \qquad \omega_i = 0, \qquad \xi = \xi(r), \qquad u = \tan\frac{\theta}{2}e^{in\phi}. \qquad (35)$$

Therefore, the boundary conditions for these new variables ξ and ω will be

$$\xi(r = 0) = \pi, \qquad \omega(r = R_0) = 0,$$

$$\omega_r(r = 0) = 0, \qquad \omega(r = R_0) = 0, \qquad (36)$$

where R_0 can be finite or infinite.

Focusing now on the solutions of the massless and massive vector models, we can compare them with the BPS Skyrme model as we have done for the baby case (see figure 5). Then, we can see that again C-compactons with the source term appear in the vector models for $\alpha = 3/2$, but not for $\alpha = 1$ where there are no solitons. This is an important issue since as we have commented above they are completely different models because of the mass hierarchy. However, in contraposition to what we found in the massive vector baby model, here we do not have a BPS solution for $\alpha = 2$, so why did we call this model BPS? The answer is that fortunately there exists a potential allowing BPS solitons both for massless and massive vectors, namely:

$$V_{BPS} = \frac{1}{4}(\xi - \cos\xi\sin\xi)^2, \qquad (37)$$

and the BPS energy calculated is

$$E = \frac{\pi^2 n^2 \lambda^2}{2} \omega(0), \tag{38}$$

which for the massless case is analytically solvable getting

$$E = \frac{\pi^4}{8\sqrt{2\sqrt{2}}} \lambda \sqrt{\frac{\lambda}{\mu}} n^{3/2}. \tag{39}$$

On the other hand, from analytical solutions for finite mass mesons we saw that in this case the energy behaves as $E \sim n$. So for both cases we have the same problem of unstable solitons for high $n^{6/5}$ we had in the baby model.

4. Conclusions

In this talk we have presented three new different BPS versions of the Skyrme model. In all of them we studied how solitons interact with the lightest fields: gauge and vector meson fields. Specially in the vector meson model we have made use of the simpler baby version of the model to help us with its generalization to the 3+1 case. In the case of the gauge field, this generalization is a more difficult task which lies outside the scope of this work.

Another important and interesting characteristic of all models is that BPS bounds have been discovered for them. In the case of the gauge model we have seen that a rich structure supports this topological bound with a superpotential equation similar to that in fake supergravity. In this case, BPS solitons exist for monotonically growing potentials and even more, all the solutions are BPS. Regarding the vector meson models, this BPS bound only exists for a specific potential: a generalized old baby potential in 2+1 dimensions and a similar to the generalized Skyrme potential for the 3+1 case.

Finally, it is necessary to emphasize that for some potentials the vector BPS models differ qualitatively from the BPS ones. It is a really important difference for the case of the standard pion mass potential, since in the BPS models we are assuming pions with infinite mass whereas in the vector ones we are adding omega mesons with finite mass, so we are changing the mass hierarchy. Therefore, if we had obtained different results, the BPS model would not be suitable from the phenomenological point of view.

Acknowledgements

The authors acknowledge financial support from the Ministry of Education, Culture and Sports, Spain (grant FPA2008-01177), the Xunta de Galicia (grant INCITE09.296.

035PR and Consellería de Educación), the Spanish Consolider-Ingenio 2010 Programme CPAN (CSD2007-00042), and FEDER. CN thanks the Spanish Ministery of Education, Culture and Sports for financial support (grant FPU AP2010-5772). Further, AW was supported by polish NCN grant 2011/01/B/ST2/00464.

References

[1] T. Skyrme, "A Nonlinear field theory", *Proc. Roy. Soc. Lond.* **A260** (1961), 127.

[2] T. Skyrme, "A Unified Field Theory of Mesons and Baryons", *Nucl. Phys.* **31** (1962), 556.

[3] T. Skyrme, "Kinks and the Dirac equation", *J. Math. Phys.* **12** (1971), 1735.

[4] B. Piette, B. Schroers, W. Zakrzewski, "Multi - solitons in a two-dimensional Skyrme model", *Z. Phys.* **C65** (1995), 165.

[5] B. Piette, B. Schroers, W. Zakrzewski, "Dynamics of baby skyrmions", *Nucl. Phys.* **B439** (1995), 205.

[6] C. Adam, J. Sanchez-Guillen, A. Wereszczynski, "A Skyrme-type proposal for baryonic matter", *Phys. Lett.* **B691** (2010), 105.

[7] C. Adam, J. Sanchez-Guillen, A. Wereszczynski, "A BPS Skyrme model and baryons at large N_c", *Phys. Rev.* **D82** (2010), 085015.

[8] E. Bonenfant, L. Marleau, "Nuclei as near BPS-Skyrmions", *Phys. Rev.* **D82** (2010), 054023.

[9] E. Bonenfant, L. Harbour, L. Marleau, "Near-BPS Skyrmions: Non-shell configurations and Coulomb effects", *Phys. Rev.* **D85** (2012), 114045.

[10] C. Adam, C. Naya, et al. "The gauged BPS baby Skyrme model", *Phys. Rev.* **D86** (2012), 045010.

[11] M. Trigiante, T. Van Riet, B. Vercnocke, "Fake supersymmetry versus Hamilton-Jacobi", *JHEP* **1205** (2012), 078.

[12] C. Adam, C. Naya, et al. "The vector BPS baby Skyrme model", *Phys. Rev.* **D86** (2012), 045015.

[13] C. Adam, C. Naya, et al. "The vector BPS Skyrme model", *Phys. Rev.* **D86** (2012), 085001.

The internal diamond target for the hypernuclear physics at HESR

H. Younis[1,a,b], **F. Balestra**[a], **M. Baydjanov**[a], **G Gallio**[a], **F. Iazzi**[a,b], **R. Introzzi**[b], **A. Lavagno**[a,b] **and V. Rigato**[c]

[a]Dept. of Applied Science and Technology, Politecnico di Torino,
 c.so Duca degli Abruzzi, 24 – 10129 Torino, Italy

[b]INFN – Sezione di Torino, c/o Politecnico di Torino,
 c.so Duca degli Abruzzi, 24, 10129 Torino, Italy

[c]INFN – LNL, Viale dell'Università, 2, 35020 Legnaro, Padova, Italy

E-mail: [1]hannan.younis@polito.it

Abstract

The case of the internal target for the hyper-nuclear experiment of the PANDA (antiproton Annihilation at Darmstadt) Collaboration at the HESR (High Energy Storage Ring) of FAIR (Facility for Antiproton and Ion Research) is illustrated. After a discussion of the problems arising from the interaction of a solid internal target with an antiproton beam, the design of material, shape and sizes of a target satisfying the experimental requirements is presented. The techniques used to produce a prototype of this target are illustrated. Then the results of the tests performed on the prototype for investigating the properties (purity, radiation hardness, structure modifications after shaping) are reported and the plan of the future activity is pointed out.

1. Introduction

A recent trend in High Energy Physics is to investigate the production of charged hyperons, the formation of doubly strange systems like double hyper-nuclei and of heavy quark systems like charmed mesons [1, 2]. To fulfill this goal, anti-protons and kaon beams are used. The cross sections of these reactions are in general very low. Therefore very intense projectile beams are necessary to obtain sets of data large enough to be significant from the statistical point of view. If the projectile-particles are difficult to be produced in large amount, it becomes mandatory to avoid as much

as possible to waste them. The machines, which supply extracted beams, waste all
the non-interacting projectiles, paying a high cost in case of low cross sections of the
reactions under investigation. In order to recover part of the non-interacting projectiles
the target can be inserted inside the beam pipe of a storage ring, as schematically
illustrated in Figure 1 (left) (extracted beam) and Figure 1 (right) (ring with internal
target). The goal of the experiment PANDA at FAIR is to reach high luminosity of the
antiproton beam, accumulated in HESR, using internal targets of ^{12}C for the hyper-
nuclear physics and of LH$_2$ for the meson spectroscopy. In this paper we shall discuss
the problems arising when a high-density target (^{12}C) is inserted inside the antiproton
storage ring: the antiproton bunch is rapidly consumed and the detector assembly
is overwhelmed by the annihilation products. The time structure of the anti-proton
beam has to be chosen in such a way that a high rate of hyperons is produced without
exceeding the maximum allowed background. A special technique of overlapping beam
spot and target will be used in PANDA and will be here described. Next, the design
of the thin ^{12}C target will be shown, together with its characteristics with respect to
the experimental requirements. After a first unsuccessful trial to produce a prototype
using amorphous ^{12}C deposition on a Cu circular support, a second trial, made starting
from diamond disks, was successful and a thin wire target has been obtained. Some
preliminary tests have been done on both the disks and the wire: the results of such
tests will be reported. Finally, conclusions will be drawn and an outlook on the future
work will be presented.

2. Interaction of antiprotons with internal targets at HESR

The PANDA Collaboration will produce double hypernuclei (DH) by means of antiprotons, using a technique based on a two-step process:

  (a) interaction of an antiproton with a nucleon to produce a Ξ^- hyperon.

  (b) absorption of the hyperon in another nucleus, after slowing down and stopping,
       and DH formation.

Two targets are necessary for this goal, one of which called internal or production
target is inserted inside the antiproton storage ring HESR. The internal target inter-
acts with the antiproton beam producing Ξ^- hyperons with a cross section $\sigma_{\Xi^-}^A(p_{\bar{p}})$
and background from hadronic reactions with cross section $\sigma_h^A(p_{\bar{p}})$. Moreover the
interaction produces losses in the bunch content, mainly due to Single Coulomb Scattering, Touschek effect and straggling. Let's call the cross sections of these processes
$\sigma_{scs}^z(p_{\bar{p}})$, σ_{Tou} and σ_{str} respectively. These losses of the antiproton beam at HESR
have been evaluated [3] at three energies for an internal hydrogen target of density
$4 \cdot 10^{15}$ atoms/cm^2. Since the slowing down to the rest of the Ξ^- hyperons must be
quick, due to their short lifetime, their production in a nuclear target instead of in

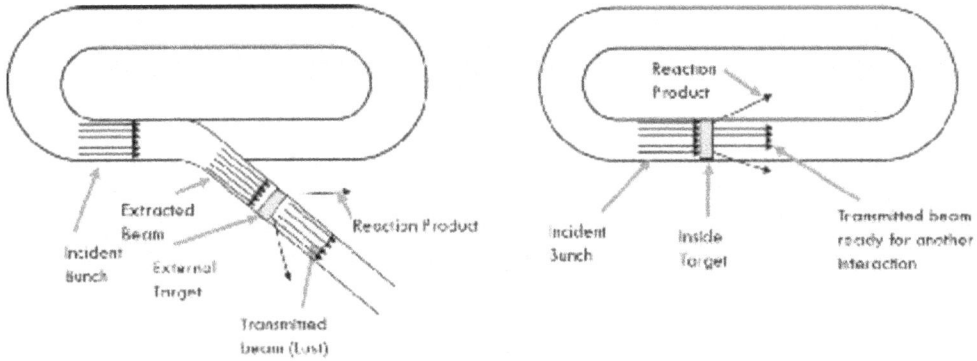

Figure 1.: (left) Scheme of an extracted beam non-interacting projectiles are lost and (right) a target inside a ring (not interacting projectiles are circulating and re-used).

a LH$_2$ target turns out to be advantageous because it strongly reduces the initial Ξ^- high energy by scattering in dense nuclear matter [4]. The use of a solid target is therefore mandatory in hyper-nuclear physics at PANDA. With anti-proton at 3 GeV/c the cross section for hyperon production in the elementary reaction antiproton-nucleon has a maximum around 2 μb. The Ξ^- production and the hadronic cross sections for a nucleus, A increase, approximately scaling with the nuclear surface, while the Single Coulomb Scattering scales with the square of the charge. Therefore at 3 GeV/c we have:

$$\sigma_{\Xi^-}^A = \sigma_{\Xi^-}^H \cdot A^{\frac{2}{3}} \approx 2 \cdot A^{\frac{2}{3}} Z \; [\mu b] \; , \tag{1}$$

$$\sigma_h^A = \sigma_h^H \cdot A^{\frac{2}{3}} \approx 75 \cdot A^{\frac{2}{3}} \; [mb] \; , \tag{2}$$

$$\sigma_{scs}^z = \frac{4\pi Z_t r_i^2}{\theta_{acc}^2} \cdot \frac{1}{\beta_o^2 \gamma_o^2} \cdot Z^2 \; [mb] \; , \tag{3}$$

where Z_t is the atomic number and A is the mass number of the target, Z is the charge of the projectile in electron charge units, β_o is the anti-proton velocity (in units of speed of light), γ_o is the corresponding Lorentz factor and r_i is the classical ion radius of the antiproton. The angle θ_{acc} is related to the acceptance and to the betatron amplitude [3]. The Touschek effect is independent on the nucleus and the straggling is weakly dependent on the target (and not increasing with A). A preliminary feasibility study of the experiment [5] indicated that the ^{12}C nucleus, optimizes the Ξ^- production and the slowing down. The sizes of the target have to be chosen in order to satisfy two requirements: the first is to produce an annihilation rate less than the maximum tolerable by the PANDA detectors (about $5 \cdot 10^6$ antiprotons/s), the second one is to consume a whole bunch at a rate not greater than the antiproton production rate of FAIR. These requirements involve the time structure of the beam at HESR, as will be discussed in the next chapter.

3. The time structure of the antiproton beam at HESR

The antiprotons produced at FAIR are accumulated in a collector ring (CR) at 3 GeV and then injected into the HESR in bunches of content I_0. At present the antiproton production rate is foreseen to be around 10^7 antiprotons/s and the bunch contents can reach at maximum 10^{10} antiprotons, produced in about 20'. After each injection a time $t_p \approx 3$–5 s (preparation time) must be spent for precooling, accelerating/decelerating (in the case of hypernuclei it's a deceleration), steering, squeezing, cooling on, cooling off and finally accelerating/decelerating. Between cooling on and cooling off there is the time t_d for the beam-target interaction. At the end of the cooling on, the target must be placed under the light of the beam spot (let r_B be the fraction of the beam spot illuminating the target) and the data taking starts. The bunch of initial content I_0 passes repeatedly through the target of thickness s and width w and after n passages the content becomes I_n:

$$I_n = I_0 \exp\left(-\frac{n}{f \cdot \tau}\right), \tag{4}$$

where the revolution frequency f and the beam lifetime τ satisfy:

$$(f \cdot \tau)^{-1} = \left[\frac{\rho \cdot N_{Av}}{A} s \cdot w \cdot r_B\right] \cdot \left[A^{\frac{2}{3}} \cdot \sigma_h^H + Z^2 \cdot \sigma_{scs}\right]. \tag{5}$$

The number $I_n^{\Xi^-}$ of Ξ^- and the number I_n^a of annihilations produced in each bunch revolution are proportional to I_n and, if the annihilation rate is less than the maximum tolerable for I_0, eq. (4) shows that it remains less for all I_n. Using a target of sizes $s \approx 3\,\mu m$ and $w \approx 100\,\mu m$, a maximum initial content $I_0 \approx 6 \cdot 10^7$ can be accepted. Due to the small lifetime the bunch is consumed in about 3 s and the total time $t_p + t_d$ for one bunch is nearly twice the data taking time. In order to strongly reduce the ratio $t_d/(t_p + t_d)$, it is possible to take advantage of the non-uniform density of the beam profile. The spot density has a radial Gaussian shape with $\sigma \approx 1\,mm$: steering continuously the beam in such a way that the spot overlaps the target only on a suitable peripheral part of its density distribution, the interacting bunch at each revolution can be tuned in order to maintain it at a nearly constant level less than I_0. Starting with $I_0 \approx 10^{10}$ and recalling that such a bunch requires about 20' to be filled, the above ratio becomes $t_d/(t_p + t_d) \approx 1$, with a great increase of the efficiency of the Ξ^- production rate.

This solution requires that a wire shaped target 3 μm thick and 100 μm wide would be made in carbon: two techniques to produce it have been explored and the results are reported in the next chapter.

4. The ^{12}C internal target at HESR

A first trial to construct a thin ^{12}C target has been made by using amorphous deposition. On a disk-shaped Cu substrate, 500 μm thick, 12 mm in diameter, a carbon film nearly 3 μm thick was deposited, purity and density were measured by means of the Back Scattering technique (BS), using a 1.5 MeV proton beam at LNL (National INFN Laboratories of Legnaro) at a scattering angle of 165°. The areal density of the film was found to be $8.3 \cdot 10^{18}$ atm/cm^2 and the ^{12}C purity level was 54%, with impurities ^{16}O (7%) and H (39%). Since the hydrogen nuclei do not slow down the hyperons, the efficiency in the Double hyperon formation should be strongly decreased. The density of the film was four times less than that of standard graphite films that means, to obtain an equivalent target thickness, the amorphous deposition would have to take four times thicker; but in such kind of films higher thickness produces higher brittleness. Moreover, five trials to cut away two rectangles inside the film, in order to shape a wire inside, by using a 30 keV Ga-ion beam, were unsuccessful, with final breakage of the wire.

For all these reasons this technique was discarded and a new one, based on diamond disks, bought from DIAMOND MATERIALS GmbH (Freiburg), has been explored. This company produced a disk-shaped diamond by Chemical Vapor Deposition on a Si plate of 15 mm diameter and a thickness of (500 ± 100) μm. An internal disk of 11 mm diameter was etched away, from this Si support thus obtaining a free standing diamond plate on a Si ring. The sizes are 15 mm outer diameter and 11 mm inner diameter and the diamond disk has a thickness of 3 ± 0.5 μm (Figure 2).

The diamond presents some important advantages with respect to the amorphous ^{12}C films. The purity is 99.9% and this avoids loss of efficiency in decelerating the hyperons. The thermal conductivity of the diamond is much better than that of the graphite: this property makes the wire suitable for dissipating the heat produced by the passage of the antiproton beam through the target. Also the mechanical resistance of the diamond is superior, allowing easier mounting and handling inside the beam pipe. The presence of a stressed region is visible (wavy diamond surface) close to the internal border of the silicon ring: the stress could be a source of brittleness and could make difficult the cutting operations. Another concern arises from the point of view of the electrical conductivity and this point will be discussed later in this chapter. The results of the sequence of tests, to which the target has been submitted, are reported in the following paragraph.

The first test was the "micro beam" test at LNL. The diamond disk has been exposed to irradiation by a 1.5 MeV proton beam in two points, P_1 close to the Si ring and P_2 close to the center of the disk, for about one and two hours, respectively. The beam spot areas were 3470×3470 μm^2 on P_1 and 870×870 μm^2 on P_2. The proton back scattering was measured during the exposure time, in the same conditions (energy

Figure 2.: Diamond disk (transparent area) stuck on to a Silicon Ring (Black Ring) The stress region is visible as wavy surface in the upper right part of the diamond circle. P_1 and P_2 indicate the irradiated points.

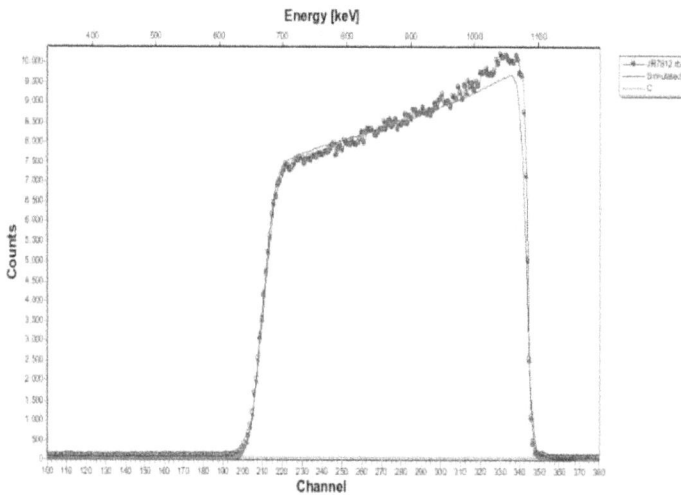

Figure 3.: Plot of the energy distribution of the back scattered protons by the wire at angle of 165° (full circles joined by a continous line named "simulated"). The curve (thin dots) corresponding to the C(natural abundance) target is superimposed and shows an excellent agreement with the data.

and angle) of the test on the amorphous film. The continuous line fits the data, shown as full dots in Figure 3, very well. The parameters of the fit are related to the type of nuclei present inside the disk, to their contents and to the areal density. The ^{12}C is present in a percentage $\approx 99.9\%$ (as expected) and the content of other nuclei is compatible with zero. The areal density is $\approx 50 \cdot 10^{18}$ atoms/cm^2, nearly 50% higher than the graphite density $\approx 34 \cdot 10^{18}$ atoms/cm^2. This result shows that the hyperon production (but also the background) will be higher than in graphite and must be taken in account in the design of the beam structure. On the other hand, the deceleration power for hyperons improves somewhat. Moreover, being the density values equal in both P_1 and P_2 with different irradiation areas and beam intensity, a first rough indication has been obtained that such an irradiation does not produce significant changes in the diamond structure.

From one of the available disks a thin wire of about 100 μm thickness and a length of about 6 cm was cut. To cut the diamond, a "Femto Edge" type LASER of 3 W power, 1064 nm wavelength and 100 fs pulse duration, has been used. The trial was successful and the picture of the wire is shown in Figure 4 (left).

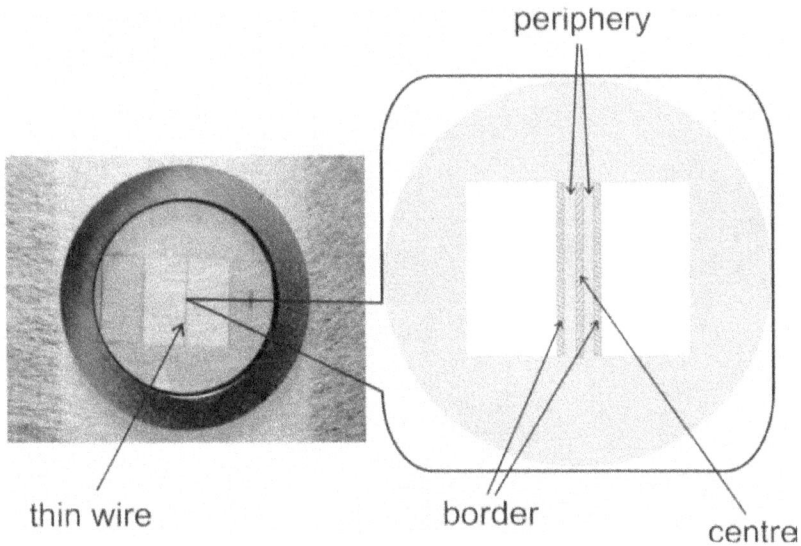

Figure 4.: (left) Picture of the diamond disk after cutting by LASER: the thin wire is visible in the centre between the two white rectangles; (right) schematic close up view of the picture, showing the three regions of the thin wire examined by Raman Spectroscopy: border (2–3 μm from the edge), periphery (around 10 μm from the edge), centre (around 50 μm from the edge).

At last, the effects of the LASER cutting on the internal structure of the wire were investigated with Raman spectroscopy. Three points have been analyzed along the width of the wire: in the center, in the periphery and in the border, as shown in the schematic detail in Figure 4 (right). The results of the Raman spectroscopy are reported in Figure 5: the bottom curve (center) shows only a clear peak in the region of the diamond (D) phase, the middle curve (periphery) shows a very slight enhancement in the region of the graphite (G) phase (with corresponding reduction in the D phase) and finally the top curve (border) shows a dominance of the G phase with respect to the D phase. One can deduce that the LASER cut seems to affect only the lateral boundary (along the cut edge) of the wire, leaving the central region unaffected. The graphitization process increases the electrical conductivity at the border. Keeping in mind the possibility that the passage of antiprotons through the target could produce electrostatic charges, a conducting channel would allow for a discharging toward the supporting frame of the ring. A special glue-paint named conducting silver paint, will be used to connect the wire to the conducting frame inside the beam pipe: The glue will cover uniformly the end part of the wire, the silicon ring and the frame, supplying a channel for electrostatic discharge. The sticking power on diamond, silicon and aluminum are under test: also the evaporation of the glue solvent as a function of the temperature is under study. The electrical conductivity of the glue is expected to be about 2 orders of magnitude less than the iron and precise measurements are in progress.

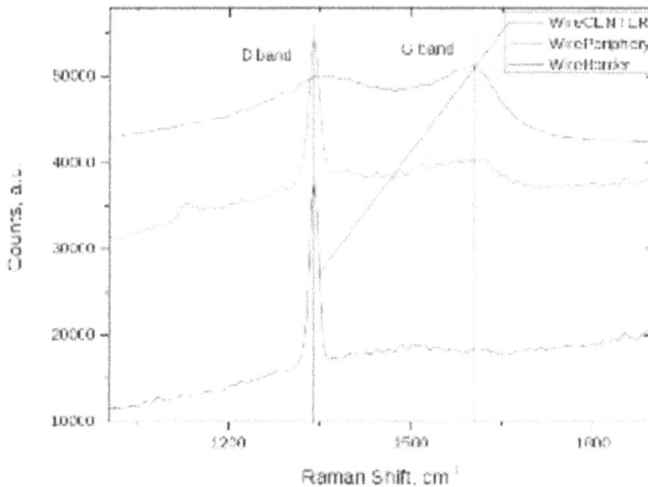

Figure 5.: Raman spectra of the central (upper curve), peripheral (middle curve) and boundary (lower curve) points along the width of the wire target, where the D band indicates the diamond phase and the G band indicates the graphite phase.

As a final remark, In the central part of the wire the diamond structure does not change and the thermal and mechanical properties remain intact, allowing a good dissipation of the heat deposited by antiprotons on the target.

5. Conclusions and future work

The presence of solid targets inside Storage Rings in high-energy physics experiments yields problems due to the alteration of the circulating projectile bunch, produced by the beam-target interaction. Therefore internal targets must satisfy severe constraints and their design and construction require special techniques. In the PANDA experiment, a diamond target, will be inserted inside the antiproton storage ring HESR at FAIR for the hyper-nuclear measurements. This target has been designed to optimize the production rate of Ξ^- hyperons without affecting appreciably the structure of the antiproton beam and to avoid an overwhelming background. A prototype of such a target has been realized starting from very thin disks of pure diamond, from which wires have been cut with technique based on a Femto-Edge LASER. The first preliminary tests gave positive results: a) purity and density of the target are suitable for the production rate and deceleration of the hyperons, b) a technique to cut the diamond by LASER has been found and c) the mechanical, thermal and electric properties of the wire shaped target satisfy the requirements of the measurements.

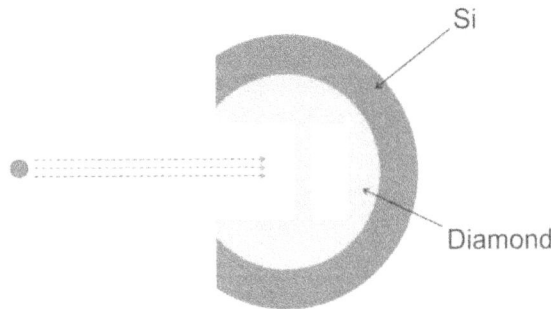

Figure 6.: Sketch of the future shape of the support of the wire diamond target. The beam spot (full circle) will be slowly steered onto the wire from the empty volume of the beam pipe.

All these tests are planned to be reported in the next future in order to get a good reliability of the parameters and of the procedures to cut and handle the wire. A future step will be the change of the supporting Si ring from the disk shape to a C-shape, as sketched in Figure 6. This new shape is required by the preliminary operations to be performed during the preparation time. The beam spot (full circle in Figure 6) is adjusted in an empty region of the beam pipe and then steered towards the wire, avoiding to overlap any material. The main problems could arise from the new

H. Younis, F. Balestra, M. Baydjanov, G. Gallio, F. Iazzi, R. Introzzi, A. Lavagno and V. Rigato

asymmetric shape of the ring, which modifies the structure of the internal stresses inside the diamond. The technique to cut the new geometry is under study.

In conclusion, the project of inserting a solid target inside the anti-proton ring at FAIR seems feasible. This solution will allow using a great part of the transmitted beam, (which would be wasted in the extracted beam machines) and could be used in future also by other facilities and other projectiles.

References

[1] K. Szymanska, F. Iazzi, "Production of the double hypernuclei with antiprotons at PANDA", *Acta Phys. Polon.* **B41** (2010), 285.

[2] K. Tanida, "Towards more exoticness – X-ray spectroscopy of Ξ atoms at J-PARC", *Hyperfine Interactions* **193** (1-3 2009), 81.

[3] A. Lehrach et al., "Beam performance and luminosity limitations in the high-energy storage ring (HESR)", *Nucl. Instrum. Meth.* **A561**.2 (2006), 289.

[4] F. Iazzi, "Doubly strange hypernuclei at PANDA", *Few-Body Systems* **43** (2008), 97.

[5] F. Ferro et al., "Ξ^- production in antiproton nucleus collisions at 3 GeV/c", *Nucl. Phys.* **A789** (2007), 209.

Going deep into supercooled water

Valentino Bianco[a], **Oriol Vilanova**[b] **and Giancarlo Franzese**[c]

Departament de Física Fonamental, Universitat de Barcelona,
C/ Martí i Franquès 1, 08028 Barcelona, Spain

E-mail: [a]vbianco@ffn.ub.edu, [b]oriol.vilanova@gmail.com, [c]gfranzese@ub.edu

Abstract

Water is probably tho most known and less understood liquid because of its anomalous properties. Many anomalies enhance when liquid water is cooled below the melting line, bringing water in a supercooled metastable state. Here, after a general introduction about what is anomalous about water, we introduce an Hamiltonian coarse-grain model for a water mono-layer confined between hydrophobic plates. We study the model using Monte Carlo simulation in NPT ensemble, exploring extensively the phase diagram at low temperature. By finite size scaling, we find a liquid-liquid first order phase transition between two metastable phases, low-density liquid (LDL) and high-density liquid (HDL), ending in a critical point (LLCP). The critical region is characterized by an increase in the thermodynamic response functions, like specific heat and compressibility. We show that the LLCP belongs to the universality class of the two-dimensional (2D) Ising model in the limit of infinite walls. Next, we modify the model in order to study the limit of stability of the liquid phase with respect to the crystal phases. We find that the model has a crystal-crystal phase transition and that the LLCP is stable with respect to the liquid-crystal phase transition depending on the relative strength of the three-body interaction with respect to the rest of many-body interactions.

1. An overview on water

Water is undoubtedly the most important liquid for life. It influences the human race in a great number of ways, going from medicine to technology. More than 60% of the human body is composed by water. Almost all biological processes need water to work correctly. Water is largely used in modern technologies and industries. Climate changes are strikingly related to water. Among all substances in nature, it is the only one occurring in solid, liquid and vapor phases at ambient conditions. The list could go on with many other aspects water is involved in.

Why do we care about water in science? Despite the fact that liquid water is so common in nature, it has a large number (overcoming sixty!) of anomalous properties which differentiate it from other liquids. Probably the most known anomaly of water is the presence of a maximum in density at finite temperature, greater than the melting temperature, leading to a crystal that is less dense than the liquid. Moreover, water is a strongly metastable liquid, indeed it can exist as a supercooled liquid at temperatures far below the melting temperature. The lowest measured temperature for supercooled liquid water is about -92 °C at 2 kbar [1]. In general we say that a mono-component substance exhibits an anomalous property if it behaves differently with respect to argon-like liquids. In particular we have a thermodynamic anomaly, dynamic anomaly or structural anomaly whenever a thermodynamic, dynamic or structural function of the system shows a non-monotonic trend upon changing the temperature T or the pressure P. Many liquid substances are found to show anomalous properties, among these we find silica (SIO_2) [2–4], silicon (Si) [5, 6], selenium (Se) [7], phosphorus (P) [8–10] and water [11].

In the following we will review some anomalous properties of water.

1.1. Thermodynamic anomalies

Water displays a maximum in density in the liquid phase, related to a negative thermal expansion coefficient, and implying that the liquid expands upon cooling or shrinks upon heating. The presence of a temperature of maximum density (TMD) for water is known since centuries: the maximum density of water is at $\approx 4\,°C$ at atmospheric pressure. For this reason the thermal expansivity $\alpha_P \equiv -(1/V)(\partial V/\partial T)_P$, where V is the volume, vanishes at $T \approx 4\,°C$ and its absolute value increases as a power law below 4 °C [12].

Other anomalous thermodynamic functions of water are the isothermal compressibility $K_T \equiv -(1/V)(\partial V/\partial P)_T$ [13] and the specific heat $C_P \equiv (\partial H/\partial T)_P$ [14], where H is the enthalpy of the system, respectively related to fluctuations in volume and entropy. For a normal liquid K_T and C_P tend to 0 upon cooling, while in the case of water they reach a minimum value at a certain temperature ($T \approx 46\,°C$ for K_T and $T \approx 35\,°C$ for C_P) and then increase for decreasing T.

The presence of a TMD at constant pressure is revealed also in othe substances, by experiments in Ga [15], Bi [16], Te [17], S [18, 19], Be, Mg, Ca, Sr, Ba [20], SiO_2, P, Se, Ce, Cs, Rb, Co, Ge [11] and by simulations on SiO_2 [21–23], S [6], BeF_2 [21]. As for the case of water, also for these substances the TMD implies anomalies in thermodynamic response functions.

1.2. Dynamic anomalies

An example of a dynamic anomaly is the non-monotonic behavior of the diffusion constant D as function of P. In a normal liquid D decreases when the density or the pressure are increased. Anomalous liquids, instead, are characterized by a region of the phase diagram where D increases upon increasing the pressure at constant temperature. In the case of water, experiments show that the normal behavior of D is restored only at pressures higher than $P \approx 1.1\,$kbar at 283 K [24].

1.3. Structural anomalies and polymorphism

A typical structural anomaly is the non-monotonic behavior of structural order parameters of the system as a function of T and P. Normal liquids tend to become more structured when compressed. The molecules adopt preferential separation and a certain orientational order. This ordering can be described by two order parameters: a translational order parameter and an orientational order parameter. The higher is the value of these parameters, the higher is the order of the system. Therefore, for normal liquids these parameters increase with increasing pressure or density at constant temperature. Anomalous liquids, instead, show a region where the system becomes more disordered as the pressure increases, leading to lower values of the structural order parameters. This is emphasized in molecular dynamics simulations for water [25] and for silica [23].

The presence of different solid structures is often related to anomalous properties of the substance. By definition a substance is polymorphic if it has several crystalline phases, and is polyamorphic if has several glassy or liquid phases. An example of a polymorph is water, with at least 17 crystalline phases [26–31], some of them stable only at high pressure. Another example is carbon with graphite, graphene and diamond. Evidences of polyamorphism in the liquid state have been observed in phosphorous [8–10], triphenyl phosphite [32–34] and in yttrium oxide–aluminum oxide melts [35]. At low temperature and low pressure water forms a low-density amorphous (LDA) ice [36]. Upon increasing the pressure it transforms from low-density (LDA) to high-density amorphous (HDA) ice [26], and upon further increasing of the pressure from HDA to very-high-density amorphous (VHDA) [37]. As we discuss later, the presence of several amorphous states could indicate the presence of a liquid-liquid phase transition.

1.4. Origin of water anomalies

It is known that most of the anomalies of water become more pronounced in the supercooled liquid phase.

Four scenarios have been proposed to explain the origin of such anomalies. The *stability limit* scenario [38] hypothesizes that the limit of stability of superheated liquid water merges with the limit of stretched and supercooled water, giving rise to a single locus in the P–T plane, with positive slope at high T and negative slope at low T. The reentrant behavior of this locus would be consistent with the anomalies of water observed at higher T. As discussed by Debenedetti, thermodynamic inconsistency challenges this scenario [39].

The *liquid-liquid critical point* (LLCP) scenario [40] supposes a first order phase transition in the supercooled region between two metastable liquids at different densities: the low-density liquid (LDL) at low P and T, and the high-density liquid (HDL) at high P and T. The phase transition line has a negative slope in the P–T plane and ends in a critical point. Numerical simulations for several models are consistent with this scenario [40–48].

The *singularity-free* scenario [49] focuses on the anticorrelation between entropy and volume as cause of the large increase of response functions at low T and hypothesizes no HB cooperativity. The scenario predicts lines of maxima in the P–T plane for the response functions, similar to those observed in the LLCP scenario, but shows no singularity for $T > 0$.

The *critical-point-free* scenario [50] hypothesizes an order-disorder transition, with a possible weak discontinuity of density, that extends to $P < 0$ and reaches the supercooled limit of stability of liquid water. This scenario would effectively predicts no critical point and a behavior for the limit of stability of liquid water as in the stability limit scenario.

To date, eperiments for bulk liquid water able to discriminate between various models are not feasible yet: the experimental resolution time is still bigger with respect the homogeneus nucleation time and the system inevitably crystallize. As we will see in the next section, the confinement in nano-structures or nano-pores is, in some cases, a way to prevent the crystal formation.

2. A Hamiltonian model for confined water

A variety of statistical models have been proposed in order to reproduce the main features of liquid water, including the anomalies. From a general stand point, different models can be classified as isotropic models or as models with directional interactions. Water molecules are often treated as rigid and the water-water interaction is modeled with isotropic potentials (like the Lennard-Jones potential) due to a specific distribution of charges on the molecule. These pair potentials reproduce water anomalies with fair agreement, but do not succeed in reproducing all the properties. For example, many of them fail in reproducing the crystal phases of water. However, the real problem with these models is that they are computationally expensive due to the long range Coulomb interaction.

This problem is particular relevant in simulations of biological processes where macro-molecules are surrounded by milions of water molecules. To overcome this problem a possible way is to consider coarse-grained models for water [51–56]. In particular, these models can be used to study nano-confined water in extreme conditions and to compare with experiments.

The study of nanoconfined water is of great interest for applications in nanotechnology and nanoscience [57]. The confinement of water in quasi-one or two dimensions (2D) is leading to the discovery of new and controversial phenomena in experiments [57–60] and simulations [58, 61–64]. Nanoconfinement, both in hydrophilic and hydrophobic materials, can keep water in the liquid phase at temperatures as low as 130 K at ambient pressure [59, 60]. At these temperatures T and pressures P experiments cannot probe liquid water in the bulk, because water freezes faster than the minimum observation time of usual techniques, resulting in an experimental "no man's land" [30]. Nevertheless, new kind of experiments and numerical simulations can access this region, revealing interesting phenomena in the metastable state [40]. Here, we will describe in detail the model proposed by Franzese and Stanley in 2002 [52–56, 65, 66].

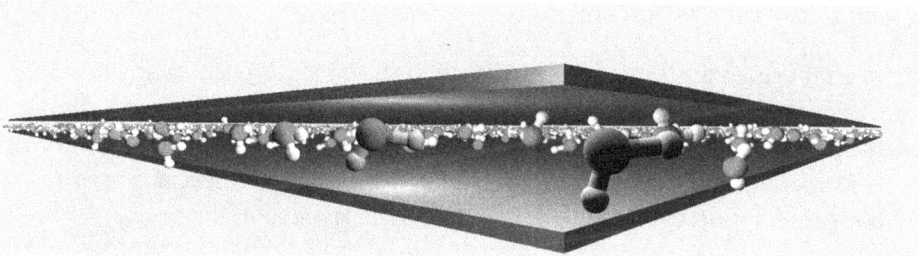

Figure 1.: Scheme of water confined between hydrophobic plates of size L. The distance between plates is $h \approx 0.5$ nm.

The simplest approximation of nano-confined water is a monolayer of water confined between two plates [67] (see fig. 2). We coarse grain the spatial distribution of the molecules dividing the system into cells, each one occupied at most by one water molecule. For each cell i we associate an occupation number $n_i = 1$ if there is a molecule in the cell (liquid-like configuration), otherwise $n_i = 0$ (gas-like configuration). The minimum volume of a cell corresponds to the hard core volume v_0 of a water molecule. Isotropic and long-range interaction (van der Waals attraction and hard core repulsion) in the system are represented by the Lennard-Jones potential

$$\mathcal{H}_{VW} \equiv -\sum_{ij} \epsilon \left[\left(\frac{r_0}{r_{ij}} \right)^{12} - \left(\frac{r_0}{r_{ij}} \right)^6 \right], \tag{1}$$

where r_{ij} is the distance between molecules i and j and the sum is performed over all the neighboring molecules up to a cutoff distance at about twenty shells. This term depends only on the relative distance between two molecules and represents the isotropic part of the interaction.

In order to describe the HB interaction, which is directional, we introduce the variables $\sigma_{ij} = 1, \ldots, q$ [49] for each occupied site i facing the cell j. Assuming that a water molecule can form up to four HBs, we fix to four the number of variables σ_{ij} for each cell. Variables σ_{ij} are introduced to account for the number of bonding configurations accessible to a water molecule. The state of a water molecule is completely determined by the values assumed by the four variables σ_{ij}. The condition for two first neighbor molecules to form a HB is $\sigma_{ij} = \sigma_{ji}$. To take into account correctly the entropy loss due to the formation of a new HB, we consider that a HB is broken if the angle $\widehat{HOO}$ deviates more than $30°$ from the linear bond. Therefore, we consider $q = 180°/30° = 6$ states, where only $1/6$ corresponds to the formation of a HB. This covalent HB interaction is represented by the Hamiltonian term

$$\mathcal{H}_{HB} \equiv -J \sum_{\langle ij \rangle} n_i n_j \delta_{\sigma_{ij}, \sigma_{ji}},$$ (2)

where $J > 0$ represent the energy gained per HB, the sum is over nearest neighbors cells, and $\delta_{ab} = 1$ if $a = b$, 0 otherwise.

Experimental evidences show that the distribution of angles $\widehat{OOO}$ changes with T and becomes sharper and sharper with the decrease of T, approaching the distribution corresponding to tetrahedral arrangement [68]. Therefore, there is a correlation among the HBs formed by the same molecule. Hence, we introduce a term representing the many-body interaction between the HBs of a single molecule,

$$\mathcal{H}_{Coop} \equiv -J_\sigma \sum_i \sum_{(k,l)_i} \delta_{\sigma_{ik}\sigma_{il}},$$ (3)

where $J_\sigma > 0$ is the characteristic energy of this cooperative component.

The formation of a HB leads to an open structure that induces a local increase of volume per molecule. This effect is incorporated in the model by considering that the total volume of the system depends linearly on the number of HBs. So the volume change is

$$V \equiv V_0 + N_{HB} v_{HB},$$ (4)

where v_{HB} is the increment due to the HB, and $V_0 \equiv N v_0$ for N water molecules.

The total enthalpy of the system is

$$H \equiv U + \mathcal{H}_{HB} + \mathcal{H}_{Coop} + PV = U - (J - P v_{HB})N_{HB} - J N_\sigma + PV_0,$$ (5)

where the total number of HB is

$$N_{\text{HB}} \equiv \sum_{\langle ij \rangle} n_i n_j \delta_{\sigma_{ij},\sigma_{ji}} , \tag{6}$$

and

$$N_\sigma \equiv \sum_i \sum_{(k,l)_i} \delta_{\sigma_{ik}\sigma_{il}} , \tag{7}$$

is the total number of HBs optimizing the cooperative interaction [52–56, 65, 67, 69, 70].

3. Thermodynamics of confined water. Phase diagram of supercooled water

We simulate this model using an efficient cluster algorithm [69, 71] that allows us to equilibrate the system very fast in the supercooled region of the phase diagram, where water exhibits glassy dynamics.

In order to characterize the phase diagram of liquid water we simulate $\approx 10^4$ different thermodynamic states for a wide range of T and P (that in real units are as low as $T = 120\,\text{K}$ and as high as $P = 0.4\,\text{GPa}$), with statistics of 5×10^6 independent calculations for systems with $N = 2500,\dots,40000$ water molecules [56, 69, 71–76]. Here, we adopt the following units: $4\epsilon/k_\text{B}$ for T and $4\epsilon/v_0$ for P, with k_B the Boltzmann constant.

The phase diagram is reported in figure 3. We find the liquid-to-gas spinodal at $P < 0$ and low T, showed with a gray line, identifying the stability limit of the liquid phase with respect to the gas phase. Then we find the reentrant line of T of maximum density (TMD) along isobars, showed with a continuous black line, that approaches the spinodal, without touching it. TMD line continues in the line of T of minimum density (TminD) as in experiments [77] and other models [78, 79], represented with a dashed black line.

We calculate the isothermal compressibility $K_T(T) \equiv -(\partial \ln \langle V \rangle / \partial P)_T$ along isobars (green lines), where $\langle V \rangle$ is the average volume, the isobaric expansivity $\alpha_P(P) \equiv (\partial \ln \langle V \rangle / \partial T)_P$ (red lines) and the isobaric specific heat $C_P(P) \equiv (\partial \langle H \rangle / \partial T)_P$ along isotherms (blue lines), where $\langle H \rangle \equiv \langle \mathcal{H} \rangle + P \langle V \rangle$ is the average enthalpy. The thermodynamic equilibrium is verified by checking that the calculation of each quantity by its definition and by the fluctuation-dissipation relation converge. For each quantity we find two maxima at low P, with higher-T maxima broader and weaker than lower-T maxima [74]. The superscripts m, wM and M refer respectively to minima, weak maxima and strong maxima of the corresponding thermodynamic quantities.

We find that both extrema of thermal expansivity along isotherms, weak maxima α_P^{wM} and minima $|\alpha|_P^{\text{wM}}$, coincides within the error bars with the extrema of compressibility

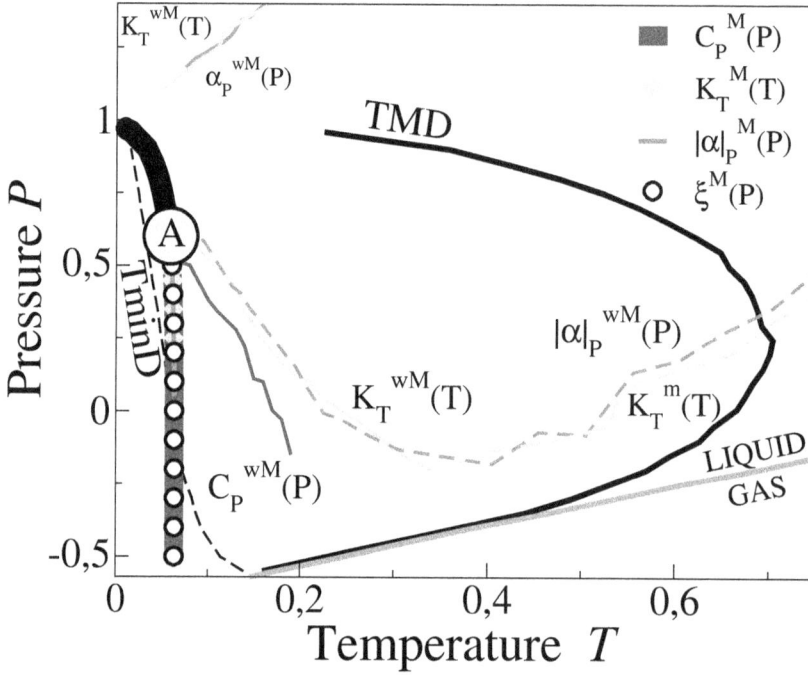

Figure 2.: Phase diagram for a monolayer of liquid water (10^4 molecules) nanoconfined between hydrophobic walls. The loci, described in the text, are marked by full lines for the maxima, dashed lines for minima, or symbols with labels nearby or in the legend. The liquid-to-gas spinodal (gray line) delimits the region of stability of the liquid with respect to the gas phase. The LLCP (large circle with A) is at the end of the LLPT (thick black) line.

along isobars, weak maxima K_T^{wM} and minima K_T^{m}, consistently with the thermodynamic relations [49]. K_T change from a maxima to a minima when $dP/dT = 0$, while α_P changes from a maxima to a minima $dP/dT = \infty$ [78, 79]. This consistency holds also for $|\alpha_P|^{\text{wM}}$ intersecting the TMD line in its turning point [49], and for the locus of weak C_P maxima (C_P^{wM}) along isotherms bending toward the turning point of the TminD line [79]. The locus of strong C_P maxima (C_P^{M}) coincides with loci of stronger maxima $|\alpha_P|^{\text{M}}$ and stronger maxima K_T^{M}.

All the loci of the maxima of the response functions converge towards the point A. Moreover, all the maxima along these loci increase in their values by approaching the point A. Bold black line represents the liquid-liquid phase transition line. Because the increase of the response functions is related to the increase of fluctuations and this is, in turn, related to the increase of the correlation length ξ, we calculate the spatial correlation function $G(r_{il}) \equiv \left\langle \sigma_{ij}(r_i)\sigma_{lk}(r_l) \right\rangle - \left\langle \sigma_{ij} \right\rangle^2$ to estimate ξ. We find that for P below the point A, $G(r)$ decays as an exponential with a characteristic correlation length ξ. For P approaching the point A, $G(r)$ is better approximated by a power-law decay with an exponential prefactor from which ξ can be extracted. At the point A, the exponential prefactor approaches a constant leaving the power-law as the dominant contribution for the decay, corresponding to ξ becoming of the order of the system size. We observe that ξ has a maximum along isobars and that the Widom line, i.e. the locus of maxima ξ^M (empty black dots), coincides with the loci of strong maxima in C_P, K_T and $|\alpha|_P$. All these observations are consistent with the identification of A with a LLCP at the end of a first-order liquid-liquid phase transition (LLPT) line in the P - T phase diagram along which the density, the energy and the entropy of the liquid are discontinuous, as discussed in previous works [54, 56, 69, 70, 72, 74, 76]. Next, we analyze in detail the LLCP.

3.1. The critical region

According to mixed-field finite-size scaling theory [80–83], a density-driven fluid-fluid phase transition is described by an order parameter $M \equiv \rho + su$, where ρ represents the leading term, u is the energy density (both in internal units) and s is the field mixing parameter. This because the order parameter distribution at the critical point must be symmetric with respect the ordered and disordered phases. Considering only the system density as order parameter will lead to a an asymmetric distribution. At the critical point the probability distribution of M is $P_N(M) \propto \tilde{p}_d(x)$, i.e. scales as an universal function $\tilde{p}_d$, characteristic of the Ising fixed point in d dimensions, of $x \equiv B(M - M_c)$, where $B \equiv a_M^{-1} N^{\beta/d\nu}$, β is the M critical exponent, ν is the ξ critical exponent, both defined by the universality class, and a_M is a non-universal system-dependent parameter. B and M_c are adjusted so that $P_N(M)$ has zero mean and unit variance.

By combining a set of 3×10^4 MC simulations for ≈ 300 state points with $0.033 \leq T \leq 0.065$ and $0.01 \leq P \leq 0.90$ with the multiple histogram reweighting method [84], and tuning s, T and P we verify that in the vicinity of the state point A the calculated $P_N(x)$ has a symmetric shape with respect to $x = 0$ (Fig. 3). We find $s = 0.25 \pm 0.03$ for our range of N. The resulting critical parameters $T_c(N)$, $P_c(N)$ and the normalization factor $B(N)$ follow in fair agreement the expected finite-size behaviors with the 2D Ising critical exponents [80–83]. From the finite-size analysis we extract the asymptotic values $T_c = 0.0597 \pm 0.0001$ and $P_c = 0.554 \pm 0.003$, consistent with the state point A. However, these fits adjust well to the data only for large N. We, therefore, perform a more systematic analysis.

For each N, we quantify the deviation of the calculated $\tilde{p}(N)$ from the expected $\tilde{p}_2$ for the 2D Ising using the Kullback-Leibler divergence [85] (showed in figure 4)

$$D_d^{\mathrm{KL}}(N) \equiv \sum_{i=1}^{n} \ln\left(\frac{\tilde{p}_{d,i}}{\tilde{p}_i(N)}\right) \tilde{p}_{d,i} \qquad (8)$$

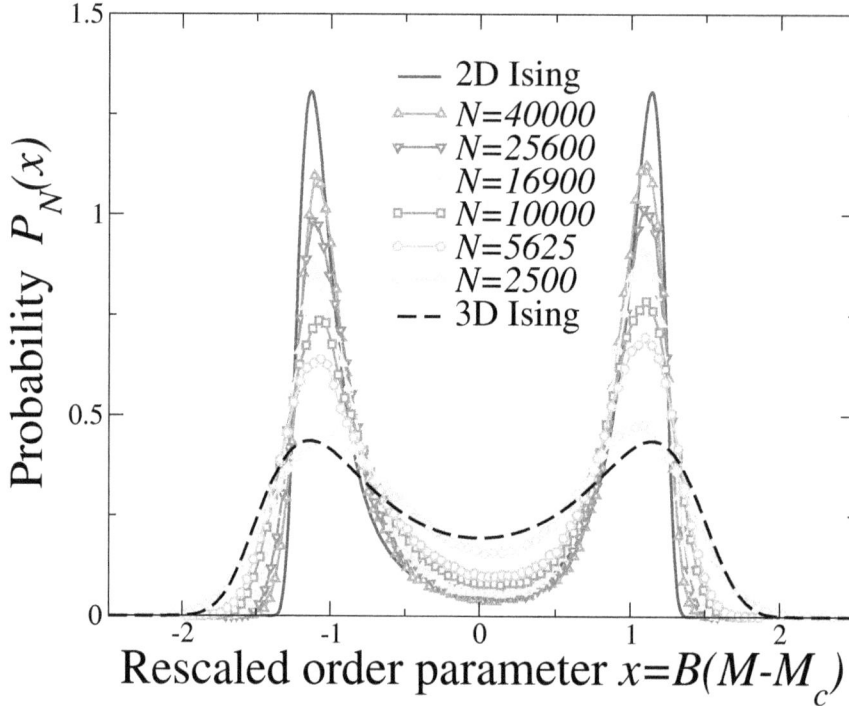

Figure 3.: The size-dependent probability distribution P_N for the rescaled order parameter x, calculated for size dependent critical temperature $T_c(N)$ and pressure $P_c(N)$, has a symmetric shape that approaches continuously (from $N = 2500$, symbols at the top at $x = 0$, to $N = 40000$, symbols at the bottom) the limiting form for the 2D Ising universality class (full line), maximizing the difference with the 3D Ising universality class (dashed line). Lines connecting the symbols are guides for the eye. Error bars are smaller than the symbols size.

of the probability distribution $\tilde{p}_i(N)$ of x_i from the theoretical value $\tilde{p}_{d,i}$ for x_i, with n total number of points for x. Furthermore, due to the behavior of data for small N in Fig. 3, we calculate the deviation from the 3D Ising $\tilde{p}_3$ [80–83].

We confirm $s \simeq 0.25$ for $\tilde{p}_2$ and find $s = 0.10 \pm 0.02$ for $\tilde{p}_3$ in our range of N. For both D_d^{KL} and W_d, with $d = 2$ and $d = 3$, we find minima that become stronger for increasing N and are always close to $T_c \simeq 0.06$ and $P_c \simeq 0.55$, i.e. at approximately constant ρ_c. We find that D_2^{KL} decreases with increasing N, vanishing for $N \to \infty$ (Fig. 4). Therefore, for an infinite monolayer confined between hydrophobic walls separated by $h \approx 0.5$ nm, the system has a LLCP that belongs to the 2D Ising universality class, consistent with our coarse-graining of the monolayer in 2D.

However, by increasing the confinement, i.e. reducing N and L at constant ρ, D_2^{KL}

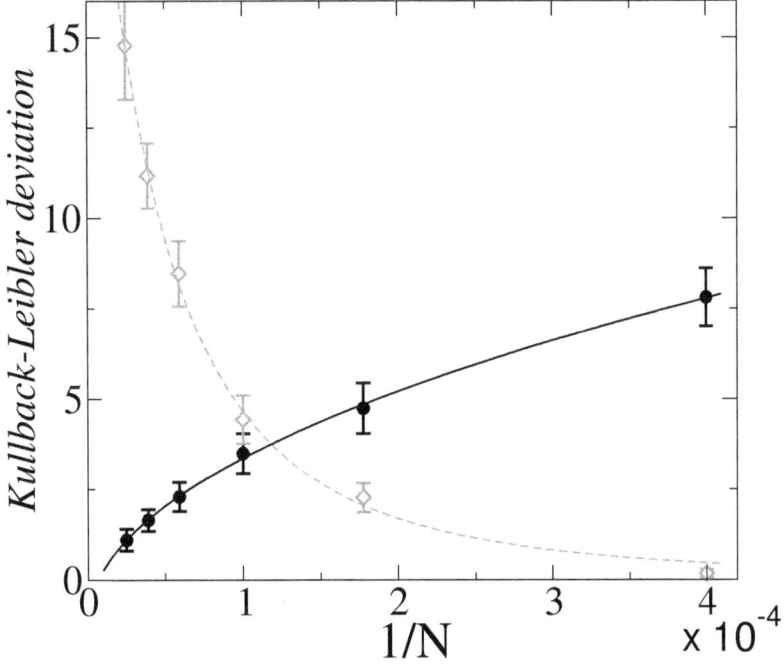

Figure 4.: Kullback-Leibler divergence $D_d^{\text{KL}}(N)$ of the calculated $\tilde{p}(N)$ from the $d = 2$ (open symbols) and $d = 3$ (closed symbols) Ising universal function $\tilde{p}_d$, as function of the inverse of the number of water molecules N at constant $\rho \simeq \rho_c$. Lines are power-law fits.

becomes larger than D_3^{KL}, respectively. For $N = 2500$ we find that D_3^{KL} has values approximately equal to those we calculate for a system ten times larger, with $D_3^{\text{KL}} \simeq 0$. In other words, by increasing the confinement of the monolayer at constant ρ, the LLCP has a behavior that approximates better the bulk [42, 47, 48, 86, 87], with a crossover between 2D and 3D-behavior occurring at $N \simeq 10^4$.

Looking at the finite-size analysis of the Gibbs free energy cost $\Delta G/(k_B T_c)$ to form an interface between the two liquids in the vicinity of the LLCP, we confirm this dimensional crossover. We calculate $\Delta G(N) \equiv -k_B T_c(N)[\ln P_N(x = 0) - \ln P_N(x_{\text{MAX}})]$, with P_N reaching a maximum at x_{MAX}. This quantity is expected to scale as $\Delta G \propto N^{\frac{d-1}{d}}$, i.e. the cost to form an interface in a d-dimensional space. We find that our data can be fitted as $N^{\frac{2}{3}}$ for small sizes and as $N^{\frac{1}{2}}$ for large sizes with a crossover around $N = 10^4$ (Fig. 5). Considering the value of the estimated ρ_c in real units ($\simeq 1\,\text{g/cm}^3$) [72, 76], the corresponding crossover wall-size is $L \simeq 25\,\text{nm}$.

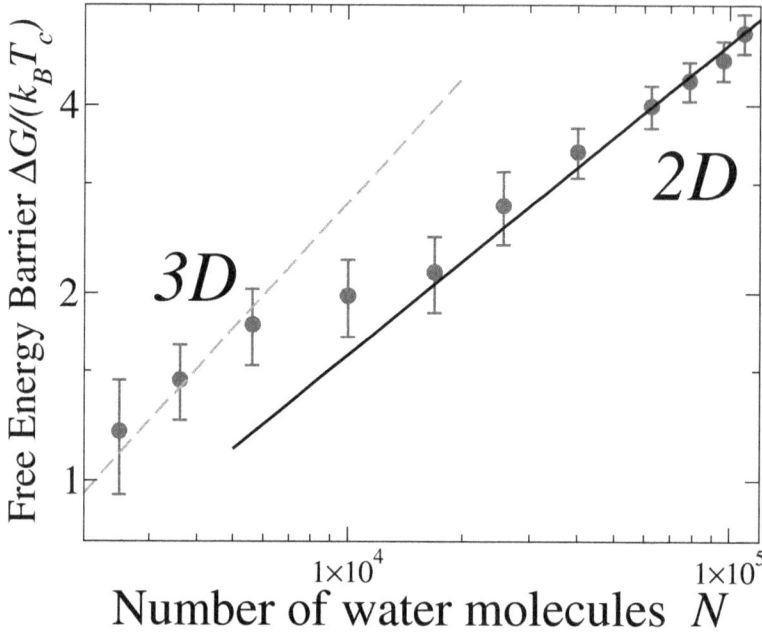

Figure 5.: The free-energy cost to form an interface between the two liquids coexisting at the LLCP scales as $\Delta G \propto N^{\frac{d-1}{d}}$ with $d = 3$ for $N < 10^4$ and $d = 2$ for $N > 10^4$.

4. Study of the Structural properties

In section 2 we introduced a coarse-grained Hamiltonian model of water that could, despite its simplicity, succesfully describe and explain a lot of properties, including some anomalies, of water. However, this description had also coarse-grained the coordinates of the molecules, being not able to reproduce the structure of the system. In this section we introduce a slight modification of the model, together with the inclusion of the explicit particle coordinates in our simulations, in order to describe structural properties, and also anomalies.

The most straightforward way of doing this was by including the following term into the Hamiltonian:

$$\mathcal{H}_{\text{Ang}} \equiv J_\theta \sum_i \sum_{(k,l)_i} \delta_{\sigma_{ik}\sigma_{ki}} \delta_{\sigma_{il}\sigma_{li}} \Delta(\theta_{k,l}^i). \tag{9}$$

This is a three-body interaction between a central molecule i and its nearest neighbors k, l, with a characteristic interaction energy J_θ. It is non-zero only if the three molecules are linked via HB, represented by the two deltas. The last element Δ is a function of the angle $\theta_{k,l}^i$ formed by the centers of the three molecules, it is purely repulsive, and with minima at angles $90°$ and $180°$. We chose these angles, not only as parameters, but as the version in 2D of the tetrahedral angles $\approx 109.5°$ that water forms in its crystalline phases, already found with MD simulations [62][61]. Its functional form is:

$$\Delta(\theta) \equiv \frac{1}{2} \left[1 + \cos(4\theta - \pi) \right] . \tag{10}$$

As long as this term depends directly on the formation of the hydrogen bonds, it will only affect the low-pressure part of the phase diagram, where the hydrogen bond network can be fully developed. The election of the magnitude of the parameter J_θ is important. It will determine the location of the melting line separating the liquid and the solid at low P. The stabilization of the solid, even at low P and low T, is still driven by the isotropic part of the Hamiltonian via the Lennard-Jones potential; so if this term is strong enough, the system will not be stable in close-packing configurations, where the preferred angles in 2D are $60°$ and $120°$, and we expect to find a new solid phase with a different symmetry. Finally, the presence of a LLPT in this model, will also interfere with this term depending on the strength of the cooperative interaction. We see that this modification in the Hamiltonian needs to be treated carefully, since all the previous terms are going to be affected by this new one.

4.1. Simulation Details and Preliminary Results

We perform Monte Carlo simulations using the Metropolis algorithm. We now have $5N + 1$ degrees of freedom: $4N$ for the σ state variables, N for the coordinates r_i of each particle and 1 for the volume of the system. We define one Monte Carlo step as $5N + 1$ trials of updating this number of degrees of freedom, chosen at random. We still keep the partition of our system in N cells. The particles are let to freely move inside its cell but not to leave it; hence each cell can be at most occupied by 1 particle. Since we expect that this last modification of the model will result in phases of different symmetries, we also let the volume change its aspect ratio. Every time a volume change is performed, we choose X or Y direction at random and change L_X or L_Y accordingly. The trial moves of the positions and the volume are performed by displacing these quantities a small random amount with a maximum value δr or δV in each case. We adapt the size of these increments to keep a constant acceptance ratio $\simeq 40\%$ [88] [89].

4.2. Radial Distribution Function

In order to study structural properties of condensed systems, the $g(r)$ is of great utility, it can help us to identify and characterize different thermodynamic phases such as liquids or solids. It is of special interest because it is directly related to the structure factor $S(k)$, which is experimentally measurable.

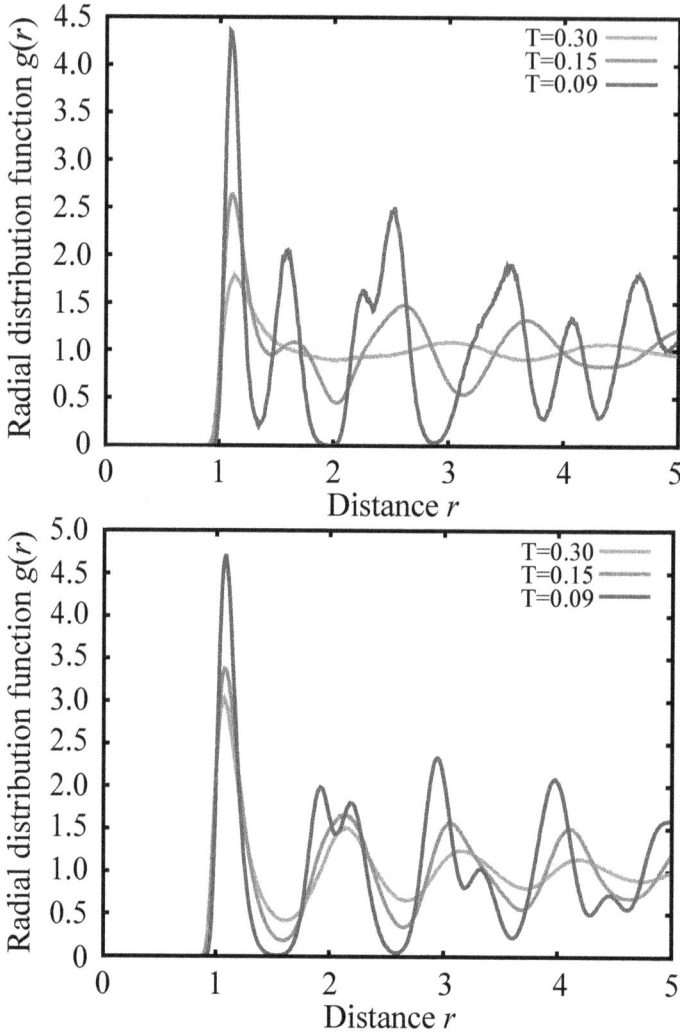

Figure 6.: Radial distribution function $g(r)$ at different temperatures and conditions of low pressure $P = 0.1$, upper plot; and at high pressure $P = 1.5$, lower plot. Each plot depicts the shapes of the $g(r)$ at different and relevant temperatures showing the passage from low-structured fluid phases at high T to highly-structured solid phases at low T. The differences depending on pressure are more evident at the solid phases, where the location of the peaks show that the system is forming two distinct crystals.

The $g(r)$, or pair-correlation function, can be defined as the probability to find a particle j at a distance r given that we have another particle i at the origin and they are separated a distance $\left| r_i - r_j \right| = r_{ij}$:

$$g(r) \equiv \frac{1}{4\pi r^2} \frac{1}{N\rho} \sum_{i=0}^{N} \sum_{i \neq j} \delta(r - r_{ij}). \tag{11}$$

In our case, we are working in two dimensions, so we actually measure the lateral pair-correlation function in X and Y directions. It is only needed to correct the prefactor and set the volume $V = L_X L_Y$ in the calculation of the number density $\rho = N/V$:

$$g_{XY}(r) \equiv \frac{1}{2\pi r} \frac{1}{N\rho} \sum_{i=0}^{N} \sum_{i \neq j} \delta(r - r_{ij}). \tag{12}$$

In Figure 4.2 we show the calculated $g(r)$ in our simulations at different conditions of temperature and pressure. For these simulations we used the parameter set: $\epsilon_{VW} = 1$, $J = 0.75$, $J_\sigma = 0.1$ and $J_\theta = 0.1$. We implemented an annealing protocol starting at high temperature and slowly decreasing T at constant P, equilibrating the system at each step. This range of temperatures show both the fluid and the solid phases at low and high pressure. The difference in the symmetry of the solid phases at low temperature is determining the shape of the $g(r)$, even in the fluid, showing that two different crystals are being formed.

4.3. Structural Order Parameters

We make use of the local bond-orientational order parameter ϕ_4, similar to that introduced by Nelson and Halperin [90], which is only function of the state of the system, and its global average Φ_4, defined as an ensemble average:

$$\Phi_4 \equiv \left\langle \frac{1}{N} \sum_{i=1}^{N} \phi_{4,i} \right\rangle = \left\langle \frac{1}{N} \sum_{i=1}^{N} \frac{1}{n_i} \sum_{j=1}^{n_i} e^{4i\theta_{ij}} \right\rangle. \tag{13}$$

The sum over each particle i is performed over its n_i nearest neighbors, where the angle θ_{ij} is the angle formed by the centers of the pair i, j with respect some reference axis. It gives information of the amount of orientational local order in the system, being $\left| \Phi_4 \right| \approx 0$ in completely random (fluid) configurations with no order, and $\left| \Phi_4 \right| \simeq 1$ at the crystalline phase consistent with this symmetry, in this case when angles are multiples of $90°$.

Next we define the translational order parameter as:

$$\Psi_G \equiv \left\langle \frac{1}{N} \sum_{i=1}^{N} \psi_{G,i} \right\rangle = \left\langle \frac{1}{N} \sum_{i=1}^{N} e^{iG \cdot r_i} \right\rangle, \tag{14}$$

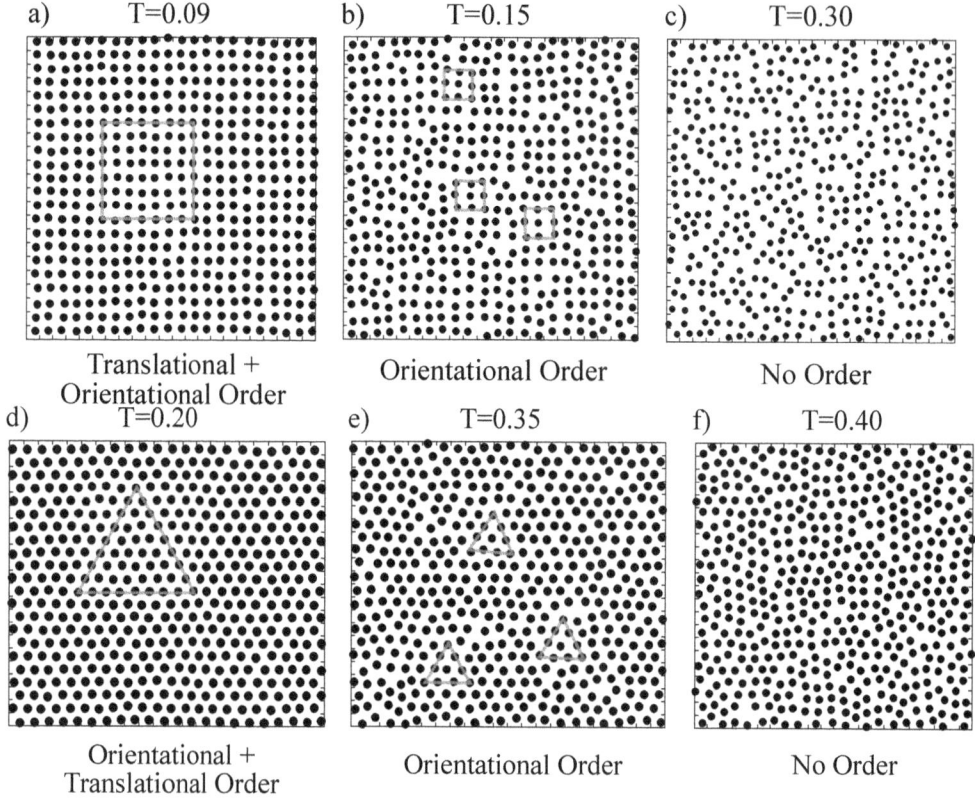

Figure 7.: Configurations of the model system at different temperatures and conditions of low pressure $P = 0.1$, plots a,b,c; and at high pressure $P = 1.5$, plots d,e,f. The orientational and translational order are quantified using the corresponding order parameters $|\Phi_4|$ and $|\Psi_G|$. Two different solid phases at low temperature are found having square and triangular symmetry at low and high pressure respectively.

where G is a vector of the reciprocal lattice corresponding to some crystal phase of reference. In our case, if we set this phase to have square symmetry, we can choose $G = \frac{2\pi}{a} e_X$, where $a = L_X / \sqrt{N}$ is the lattice spacing. This new order parameter gives information of global order in the system, and helps to identify the crystal phase, because it is $|\Psi| \simeq 1$ only when the system is highly structured [91].

In Figure 4.3, we show the plots of the configurations corresponding to the same conditions as the represented with the radial distribution functions in Figure 4.2. We characterize the different kinds of order depending on the value of the structural order parameters already defined. We also used the corresponding pair of order parameters for the triangular symmetry Φ_6 and Ψ_G with similar definitions to the square symmetry ones. The onset of each kind of order is determined when the order parameters attain a threshold value above 0.8, which is the approximate value where the $g(r)$ show important changes.

5. Conclusions

In conclusion, we review some anomalous preperties of water and report an Hamiltonian coarse-grain model of a water monolayer confined between hydrophobic parallel walls of lateral size L separated by $h \approx 0.5$ nm, well reproducing thermodynamic and structural anomalies of water. We analyze the low-T phase diagram. We identify many loci of the phase diagram, the liquid-gas spinodal, the Widom line, the extrema in thermodynamic response functions, the LLPT and the LLCP. We show that the LLCP belongs to the 2D Ising universality class only for $L/h \geq 50$. For smaller L at the same ρ, i.e. for stronger confinement, the LLCP is better described by 3D, bulk-like, critical behavior, as a consequence of the high cooperativity and low coordination number of the HB network.

We also introduced a change in the Hamiltonian that let us to study the structural properties of the system and found up to two different kinds of crystal phases with different symmetries. We characterized these phases by analyzing the radial distribution distribution function and the bond-orientational and translational order parameters of the system.

Acknowledgements

We thank M. C. Barbosa, M. Bernabei, S. V. Buldyrcv, F. Bruni, S.-H. Chen, P. Debenedetti, P. Kumar, F. Leoni, J. Luo, G. Malescio, F. Mallamace, M. I. Marqus, M. G. Mazza, A.B. de Oliveira, S. Pagnotta, D. Reguera, F. de los Santos, F. Sciortino, H. E. Stanley, F. W. Starr, K. Stokely, E. G. Strekalova and P. Vilaseca for helpful discussions. We thank for support the Spanish Ministerio de Ciencia e Innovación FIS2009-10210, the Spanish Ministerio de Economia y Competitividad Grant FIS2012-31025, the Generalitat de Catalunya Grant 2010 FI-DGR, and the European Commission Grant FP7-NMP-2010-EU-USA.

References

[1]   H. Kanno, R. J. Speedy, C. A. Angell, "Supercooling of water to -92 °C under pressure", *Science* **189**.4206 (1975), 880.

[2]   C. A. Angell, S. Borick, M. Grabow, "Glass transitions and first order liquid-metal-to-semiconductor transitions in 4-5-6 covalent systems", *J. Non-Cryst. Solids* **205-207**.96 (1996), 463.

[3]   P. H. Poole, M. Hemmati, C. A. Angell, "Comparison of Thermodynamic Properties of Simulated Liquid Silica and Water", *Phys. Rev. Lett.* **79**.12 (1997), 2281.

[4]   D. J. Lacks, "First-Order Amorphous-Amorphous Transformation in Silica", *Phys. Rev. Lett.* **84**.20 (2000), 4629.

[5]   I. Saika-Voivod, F. Sciortino, P. H. Poole, "Computer simulations of liquid silica: Equation of state and liquid–liquid phase transition", *Phys. Rev.* **E63**.1 (2000), 011202.

[6]   S. Sastry, C. Austen Angell, "Liquid-liquid phase transition in supercooled silicon", *Nature. Mat.* **2** (2003), 739.

[7]   V. V. Brazhkin, S. V. Popova, R. N. Voloshin, "High-pressure transformations in simple melts", *High Pressure Res.* **15**.5 (1997), 267.

[8]   Y. Katayama et al., "A first-order liquid-liquid phase transition in phosphorus", *Nature* **403**.6766 (2000), 170.

[9]   Y. Katayama et al., "Macroscopic Separation of Dense Fluid Phase and Liquid Phase of Phosphorus", *Science* **306**.5697 (2004), 848.

[10] G. Monaco et al., "Nature of the First-Order Phase Transition in Fluid Phosphorus at High Temperature and Pressure", *Phys. Rev. Lett.* **90**.25 (2003), 255701.

[11] P. G. Debenedetti, "Metastable Liquids. Concepts and Principles", Princeton University Press (Princeton, NJ), 1996.

[12] D. E. Hare, C. M. Sorensen, "Densities of supercooled H_2O and D_2O in 25 mu glass capillaries", *J. Chem. Phys.* **84**.9 (1986), 5085.

[13] R. J. Speedy, C. A. Angell, "Isothermal compressibility of supercooled water and evidence for a thermodynamic singularity at -45 °C", *J. Phys. Chem.* **65**.3 (1976), 851.

[14] C. A. Angell, W. J. Sichina, M. Oguni, "Heat capacity of water at extremes of supercooling and superheating", *J. Phys. Chem.* **86**.6 (1982), 998.

[15] K. K. Mon, N. W. Ashcroft, G. V. Chester, "Core polarization and the structure of simple metals", *Phys. Rev.* **B19**.10 (1979), 5103.

[16] P. Lamparter, S. Steeb, W. Knoll, "Structure of Molten Bi-Sb-alloys by Means of Neutron Diffraction", *Z. Naturf. A* **31** (1976), 90.

[17] H. Thurn, J. Ruska, "Change of bonding system in liquid SexTe1-1 alloys as shown by density measurements", *J. Non-Cryst. Solids* **22**.2 (1976), 331.

[18] G. E. Sauer, L. B. Borst, "Lambda Transition in Liquid Sulfur", *Science* **158**.3808 (1967), 1567.

[19] S. J. Kennedy, J. C. Wheeler, "On the density anomaly in sulfur at the polymerization transition", *J. Chem. Phys.* **78**.3 (1983), 1523.

[20] J.-F. Wax, R. Albaki, J.-L. Bretonnet, "Temperature dependence of the diffusion coefficient in liquid alkali metals", *Phys. Rev.* **B65**.1 (2002), 014301.

[21] C. A. Angell et al., "Water and its anomalies in perspective: tetrahedral liquids with and without liquid-liquid phase transitions. Invited Lecture", *Phys. Chem. Chem. Phys.* **2** (8 2000), 1559.

[22] R. Sharma, S. N. Chakraborty, C. Chakravarty, "Entropy, diffusivity, and structural order in liquids with waterlike anomalies", *J. Chem. Phys.* **125**.20, 204501 (2006), 204501.

[23] M. S. Shell, P. G. Debenedetti, A. Z. Panagiotopoulos, "Saddles in the Energy Landscape: Extensivity and Thermodynamic Formalism", *Phys. Rev. Lett.* **92**.3 (2004), 035506.

[24] C. A. Angell, E. D. Finch, P. Bach, "Spin–echo diffusion coefficients of water to 2380 bar and -20 °C", *J. Chem. Phys.* **65**.8 (1976), 3063.

[25] J. R. Errington, P. G. Debenedetti, "Relationship between structural order and the anomalies of liquid water", *Nature* **409** (2001), 318.

[26] O. Mishima, L. D. Calvert, E. Whalley, "An Apparently 1st-Order Transition Between 2 Amorphous Phases Of Ice Induced By Pressure", *Nature* **314**.6006 (1985), 76.

[27] O. Mishima, "Reversible first-order transition between two H_2O amorphs at ~ 0.2 GPa and ~ 135 K", *J. Chem. Phys.* **100**.8 (1994), 5910.

[28] O. Mishima, "Relationship between melting and amorphization of ice", *Nature* **384** (1996), 546.

[29] O. Mishima, Y. Suzuki, "Propagation of the polyamorphic transition of ice and the liquid-liquid critical point", *Nature* **419** (2002), 599.

[30] O. Mishima, H. E. Stanley, "The relationship between liquid, supercooled and glassy water", *Nature* **396**.6709 (1998), 329.

[31] G. Franzese, H. E. Stanley, "Understanding the Unusual Properties of Water", in *Water and Life: The Unique Properties of H_2O*, ed. by R. M. Lynden-Bell et al., CRC Press ()2010, ISBN: 978-1-4398-0356-1.

[32] R. Kurita, H. Tanaka, "Critical-Like Phenomena Associated with Liquid-Liquid Transition in a Molecular Liquid", *Science* **306**.5697 (2004), 845.

[33] H. Tanaka, R. Kurita, H. Mataki, "Liquid-Liquid Transition in the Molecular Liquid Triphenyl Phosphite", *Phys. Rev. Lett.* **92**.2 (2004), 025701.

[34] R. Kurita, H. Tanaka, "On the abundance and general nature of the liquid-liquid phase transition in molecular systems", *J. Phys. Cond. Matter* **17**.27 (2005), L293.

[35] G. N. Greaves et al., "Detection of First-Order Liquid/Liquid Phase Transitions in Yttrium Oxide-Aluminum Oxide Melts", *Science* **322**.5901 (2008), 566.

[36]   P. Brüggeller, E. Mayer, "Complete vitrification in pure liquid water and dilute aqueous solutions", *Nature* **288** (1980), 569.

[37]   J. L. Finney et al., "Structure of a New Dense Amorphous Ice", *Phys. Rev. Lett.* **89**.20 (2002), 205503.

[38]   R. J. Speedy, "Limiting forms of the thermodynamic divergences at the conjectured stability limits in superheated and supercooled water", *J. Phys. Chem.* **86**.15 (1982), 3002.

[39]   P. G. Debenedetti, "Supercooled and glassy water", *J. Phys. Cond. Matter* **15**.45 (2003), R1669.

[40]   P. H. Poole et al., "Phase-Behavior Of Metastable Water", *Nature* **360**.6402 (1992), 324.

[41]   P. H. Poole et al., "Effect of Hydrogen Bonds on the Thermodynamic Behavior of Liquid Water", *Phys. Rev. Lett.* **73**.12 (1994), 1632.

[42]   H. Tanaka, "A self-consistent phase diagram for supercooled water", *Nature* **380** (1996), 328.

[43]   H. Tanaka, "Phase behaviors of supercooled water: Reconciling a critical point of amorphous ices with spinodal instability", *J. Chem. Phys.* **105**.12 (1996), 5099.

[44]   D. Paschek, A. Rüppert, A. Geiger, "Thermodynamic and Structural Characterization of the Transformation from a Metastable Low-Density to a Very High-Density Form of Supercooled TIP4P-Ew Model Water", *Chem. Phys. Chem.* **9**.18 (2008), 2737.

[45]   Y. Liu, A. Z. Panagiotopoulos, P. G. Debenedetti, "Low-temperature fluid-phase behavior of ST2 water", *J. Chem. Phys.* **131**.10, 104508 (2009), 104508.

[46]   D. A. Fuentevilla, M. A. Anisimov, "Scaled Equation of State for Supercooled Water near the Liquid-Liquid Critical Point", *Phys. Rev. Lett.* **97**.19 (2006), 195702.

[47]   J. L. F. Abascal, C. Vega, "Widom line and the liquid-liquid critical point for the TIP4P/2005 water model", *J. Chem. Phys.* **133**.23, 234502 (2010), 234502.

[48]   K. Tobias et al., "Nanoscale Dynamics of Phase Flipping in Water near its Hypothesized Liquid-Liquid Critical Point", *Nature Sci. Rep.* **2** (2012), 474.

[49]   S. Sastry et al., "Singularity-free interpretation of the thermodynamics of supercooled water", *Phys. Rev.* **E53**.6 (1996), 6144.

[50]   C. A. Angell, "Insights into Phases of Liquid Water from Study of Its Unusual Glass-Forming Properties", *Science* **319**.5863 (2008), 582.

[51]   V. Molinero, E. B. Moore, "Water modeled as an intermediate element between carbon and silicon.", *J. Phys. Chem. B* **113**.13 (2009), 4008.

[52] G. Franzese, H. E. Stanley, "A theory for discriminating the mechanism respon-sible for the water density anomaly", *Physica A Stat. Mech.* **314**.1-4 (2002), 508.

[53] G. Franzese, H. E. Stanley, "Liquid-liquid critical point in a Hamiltonian model for water: analytic solution", *J. Phys. Cond. Matter* **14**.9 (2002), 2201.

[54] G. Franzese, M. I. Marques, H. E. Stanley, "Intramolecular coupling as a mecha-nism for a liquid-liquid phase transition", *Phys. Rev.* **E67**.1 (2003), 011103.

[55] G. Franzese, H. E. Stanley, "The Widom line of supercooled water", *J. Phys. Cond. Matter* **19**.20 (2007), 205126.

[56] P. Kumar, G. Franzese, H. E. Stanley, "Predictions of Dynamic Behavior under Pressure for Two Scenarios to Explain Water Anomalies", *Phys. Rev. Lett.* **100**.10 (2008), 105701.

[57] D. R. Paul, "Creating New Types of Carbon-Based Membranes", *Science* **335**.6067 (2012), 413.

[58] M. Whitby, N. Quirke, "Fluid flow in carbon nanotubes and nanopipes", *Na-ture Nanotechn:* **2** (2 2007), 87.

[59] Y. Zhang et al., "Density hysteresis of heavy water confined in a nanoporous silica matrix", *Proc. Nat. Acad. Sci.* **108**.30 (2011), 12206.

[60] A. K. Soper, "Density minimum in supercooled confined water", *Proc. Nat. Acad. Sci.* **108**.47 (2011), E1192.

[61] R. Zangi, A. E. Mark, "Monolayer Ice", *Phys. Rev. Lett.* **91**.2 (2003), 025502.

[62] P. Kumar et al., "Thermodynamics, structure, and dynamics of water confined between hydrophobic plates", *Phys. Rev.* **E72**.5 (2005), 051503.

[63] M. Sharma et al., "Probing Properties of Water under Confinement: Infrared Spectra", *Nano Lett.* **8**.9 (2008), 2959.

[64] R. R. Nair et al., "Unimpeded Permeation of Water Through Helium-Leak–Tight Graphene-Based Membranes", *Science* **335**.6067 (2012), 442.

[65] G. Franzese et al., "Phase transitions and dynamics of bulk and interfacial water", *J. Phys. Cond. Matter* **22**.28 (2010), 284103.

[66] V. Bianco, G. Franzese, "Critical behavior of a water monolayer under hydropho-bic confinement" (2012), arXiv:1212.2847.

[67] G. Franzese, F. de los Santos, "Dynamically Slow Processes in Supercooled Water Confined Between Hydrophobic Plates", *J. Phys. Cond. Matter* **21** (2009), 504107.

[68] M. A. Ricci, F. Bruni, A. Giuliani, "Similarities between confined and supercooled water", *Faraday Discussions* **141** (2009), 347.

[69] M. G. Mazza et al., "Cluster Monte Carlo and numerical mean field analysis for the water liquid-liquid phase transition", *Comp. Phys. Commun.* **180**.4 (2009), 497.

[70] K. Stokely et al., "Effect of hydrogen bond cooperativity on the behavior of water", *Proc. Nat. Acad. Sci.* **107** (2010), 1301.

[71] V. Bianco, S. Iskrov, G. Franzese, "Understanding the role of hydrogen bonds on water dynamics and protein stability", *J. Biol. Phys.* **38** (1 2011), 27.

[72] F. de los Santos, G. Franzese, "Understanding diffusion and density anomaly in a coarse-grained model for water confined between hydrophobic walls", *J. Phys. Chem. B* **115** (1520-5207 2011), 14311.

[73] G. Franzese, V. Bianco, S. Iskrov, "Water at Interface with Proteins", *Food Biophys.* **6** (2 2011), 186.

[74] M. G. Mazza et al., "More than one dynamic crossover in protein hydration water", *Proc. Nat. Acad. Sci.* **108**.50 (2011), 19873.

[75] E. G. Strekalova et al., "Large Decrease of Fluctuations for Supercooled Water in Hydrophobic Nanoconfinement", *Phys. Rev. Lett.* **106**.14 (2011), 145701.

[76] F. de los Santos, G. Franzese, "Relations between the diffusion anomaly and cooperative rearranging regions in a hydrophobically nanoconfined water monolayer", *Phys. Rev.* **E85** (1 2012), 010602.

[77] F. Mallamace et al., "The anomalous behavior of the density of water in the range 30 K < T < 373 K", *Proc. Nat. Acad. Sci.* **104**.47 (2007), 18387.

[78] V. V. Vasisht, S. Saw, S. Sastry, "Liquid-liquid critical point in supercooled silicon", *Nature Phys.* **7** (2011), 549.

[79] I. S.-V. Peter H Poole, F. Sciortino, "Density minimum and liquid-liquid phase transition", *J. Phys. Cond. Matter* **17**.43 (2005).

[80] A. D. Bruce, N. B. Wilding, "Scaling fields and universality of the liquid-gas critical point", *Phys. Rev. Lett.* **68** (2 1992), 193.

[81] N. B. Wilding, "Critical-point and coexistence-curve properties of the Lennard-Jones fluid: A finite-size scaling study", *Phys. Rev.* **E52** (1 1995), 602.

[82] R. Hilfer, N. B. Wilding, "Are critical finite-size scaling functions calculable from knowledge of an appropriate critical exponent?", *J. Phys. A* **28**.10 (1995), L281.

[83] N. B. Wilding, K. Binder, "Finite-size scaling for near-critical continuum fluids at constant pressure", *Physica A Stat. Mech.* **231**.4 (1996), 439.

[84] A. Z. Panagiotopoulos, "Monte Carlo methods for phase equilibria of fluids", *J. Phys. Cond. Matter* **12**.3 (2000), R25.

[85] S. Kullback, R. A. Leibler, *Ann. Math. Stat.* **22** (1951), 79.

[86] F. Sciortino, I. Saika-Voivod, P. H. Poole, "Study of the ST2 model of water close to the liquid-liquid critical point", *Phys. Chem. Chem. Phys.* **13** (44 2011), 19759.

[87] Y. Liu et al., "Liquid-liquid transition in ST2 water", *J. Chem. Phys.* **137**.21, 214505 (2012), 214505.

[88] D. Bouzida, S. Kumar, R. H. Swendsen, "Efficient Monte Carlo methods for the computer simulation of biological molecules", *Phys. Rev.* **A45** (12 1992), 8894.

[89] I. R. McDonald, "NpT-ensemble Monte Carlo calculations for binary liquid mixtures", *Mol. Phys.* **100**.1 (2002), 95–105.

[90] B. I. Halperin, D. R. Nelson, "Theory of Two-Dimensional Melting", *Phys. Rev. Lett.* **41** (2 1978), 121.

[91] K. Wierschem, E. Manousakis, "Simulation of melting of two-dimensional Lennard-Jones solids", *Phys. Rev.* **B83** (21 2011), 214108.

III

Supersymmetry and Supergravity

Supersymmetric BCS superconductivity

Alejandro Barranco[1] and Jorge G. Russo[2]

Departament d'Estructura i Constituents de la Matèria, Universitat de Barcelona, C/ Martí Franquès 1, 08028 Barcelona, Spain

E-mail: [1]alejandro@ecm.ub.edu, [2]jrusso@ecm.ub.edu

Abstract

We implement relativistic BCS superconductivity in $\mathcal{N} = 1$ supersymmetric field theories with a $U(1)_R$ symmetry. The simplest model contains two chiral superfields with a Kähler potential modified by quartic terms. The superconducting phase transition turns out to be first order, due to the scalar contribution to the one-loop potential.

1. Introduction

In a supersymmetric theory, fermions and scalars are combined with very specific couplings, so that these theories are more stable under radiative corrections due to one-loop cancellations between fermionic and bosonic contributions. For this reason, understanding how superconducting transitions can take place in supersymmetric field theories in full detail is of great interest, in particular, to clarify how supersymmetric theories react upon the introduction of a chemical potential. Other motivations include providing a field-theoretical understanding of the possible mechanisms underlying holographic superconductivity, and possible applications to real-world condensed matter systems containing fermion and scalar quasiparticle excitations.

Here we present a brief review of [1], where a model of supersymmetric BCS super-conductivity was proposed. After reviewing relativistic BCS theory [2, 3] in section 2, we will see that BCS requires the introduction of a chemical potential for fermions and in supersymmetric theories this leads to some problems. Consistency demands that this chemical potential be coupled to a (non-anomalous) U(1) current. For a baryonic $U(1)_B$ symmetry, in supersymmetric theories, this can only be done in a consistent way by simultaneously introducing the same chemical potential for the scalars. But charged scalar fields in the presence of chemical potential can run into problems of Bose-Einstein (BE) condensation when the chemical potential becomes greater than the mass. Therefore, the one-loop potential becomes ill-defined due to divergences. Adding a mass term to the superpotential does not circumvent this problem because the requirement of existence of Fermi surfaces, due to the relations between mass parameters implied by supersymmetry, is always correlated to the appearance of BE condensation. All these problems are explained in section 3.1. In section 3.2, we will evade this problem by coupling the chemical potential to a $U(1)_R$ current and considering a model where the light scalars have vanishing $U(1)_R$ charge. We end in section 4 with some conclusions. We will follow the notation of [4].

2. Review of relativistic BCS theory

It is useful to briefly review the main features of relativistic BCS theory. Here we will work with global $U(1)$ symmetries, so in this sense we will be discussing superfluidity, although transport properties are similar in both cases. In relativistic BCS theory, one has the effective Lagrangian [3]

$$\mathcal{L} = \frac{i}{2}\left(\bar{\psi}\gamma^\mu\partial_\mu\psi - \partial_\mu\bar{\psi}\gamma^\mu\psi\right) - m\bar{\psi}\psi + \mu\psi^\dagger\psi + \frac{g^2}{2}\left(\bar{\psi}_c\gamma_5\psi\right)^\dagger\left(\bar{\psi}_c\gamma_5\psi\right). \quad (1)$$

The $U(1)$ symmetry ensures fermion number conservation, which allows for the introduction of a chemical potential in the usual way. The Lagrangian is not renormalizable, the four-fermion interaction typically represents an irrelevant operator, but the dynamics of BCS superconductivity is such that, for fermions which are close to the Fermi surface, this attractive four-fermion interaction becomes strong. At weak coupling the scaling dimension of the fermionic fields must be very close to that of a free field, 3/2. Hence, on dimensional grounds, the interaction term is irrelevant in the IR. Naively, it would seem that this theory cannot lead to any interesting IR physics. The phenomenon that actually takes place is explained in [5]. The key observation is that in the presence of a chemical potential there is a Fermi surface which can change the naive scaling dimensions for the operators in such a way that the otherwise irrelevant interaction becomes indeed marginal. This is the seed for the possibility of non-trivial IR physics, such as superconductivity.

Technically, to study this system, one considers the Euclidean theory at finite temperature and performs a Hubbard-Stratonovich transformation. One introduces the auxiliary field $\Delta(x)$ and the Lagrangian is then replaced by

$$\mathcal{L}_E = \frac{1}{2}\left(\psi^\dagger\partial_\tau\psi - \partial_\tau\psi^\dagger\psi\right) - \frac{i}{2}\left(\bar{\psi}\gamma^i\partial_i\psi - \partial_i\bar{\psi}\gamma^i\psi\right) + m\bar{\psi}\psi - \mu\psi^\dagger\psi$$
$$+ \frac{1}{2g^2}|\Delta|^2 - \frac{1}{2}\left[\Delta^\dagger\left(\bar{\psi}_c\gamma_5\psi\right) + \Delta\left(\bar{\psi}_c\gamma_5\psi\right)^\dagger\right]. \quad (2)$$

The Lagrangian becomes quadratic in the fermions, which can now be integrated out explicitly, giving rise to an effective potential for Δ. The fermion energy eigenvalues are

$$\omega_\pm = \sqrt{(\omega_0(p) \pm \mu)^2 + |\Delta|^2}, \qquad \omega_0 \equiv \sqrt{p^2 + m^2}, \tag{3}$$

where $\pm$ stands for particles and antiparticles. The one-loop effective potential Ω (fig. 1a) is then obtained as usual by adding to the classical potential, $\frac{1}{2g^2}|\Delta|^2$, the thermal contribution

$$-\frac{2}{\beta} \int \frac{d^3p}{(2\pi)^3} \left(\log(1 + e^{-\beta \omega_-(p)}) + \log(1 + e^{-\beta \omega_+(p)}) \right),$$

plus a (Coleman-Weinberg) contribution that survives at zero temperature,

$$\int \frac{d^3p}{(2\pi)^3} \left(2\omega_0(p) - \omega_-(p) - \omega_+(p) \right). \tag{4}$$

Ω is identified with the thermodynamic potential of the grand canonical ensemble. The integral over momentum for this contribution is divergent. Since the theory is not renormalizable, one must restrict to energies below a cutoff Λ ("Debye" energy). The cutoff as usual represents the energy where new physics emerges. At low temperatures, the dominant contribution then arises from frequencies ω_0 near μ and the contribution of the antiparticle –represented by the terms with ω_+– can be neglected (we assume $\mu > 0$; if $\mu < 0$ it is the antiparticle contribution the dominant one). Since $p^2 > 0$, the existence of a Fermi surface at finite momentum p_F requires $\mu > m$, where p_F is defined by the condition $\sqrt{p_F^2 + m^2} = \mu$. As a result, the system has a Fermi energy represented by the chemical potential μ. If $\Delta = 0$, this represents the Fermi energy in the usual sense, at zero temperature fermions would occupy energy levels with $\omega_0(p) < \mu$. However, in this system, Δ is spontaneously turned on below some critical temperature. When Δ is not equal to zero, there is a fermion condensate and the energy eigenvalues $\omega_\pm(p)$ do not vanish at any value of momentum. At low temperatures, the dominant contributions come from the region where $\omega_-(p)$ has a minimum value. The location in momentum space of this minimum value defines the concept of Fermi surface in more general situations. For this system, this still occurs at $\omega_0(p) = \mu$.

The instability leading to $\Delta \neq 0$ and thus to the formation of the fermion condensate appears when the coefficient of the $O(\Delta^2)$ term in the complete expression for the one-loop effective potential changes sign. The gap equation $\Delta = \Delta(T)$ is obtained by differentiating the one-loop effective potential with respect to Δ. We find

$$1 = \frac{g^2}{2\pi^2} \int_0^\Lambda dp\, p^2 \left(\frac{\tanh\left(\frac{1}{2}\beta\omega_-(p, \Delta)\right)}{\omega_-(p, \Delta)} + \frac{\tanh\left(\frac{1}{2}\beta\omega_+(p, \Delta)\right)}{\omega_+(p, \Delta)} \right). \tag{5}$$

This gives the usual critical curve for a second-order phase transition for the order parameter Δ as a function of the temperature (fig. 1b).

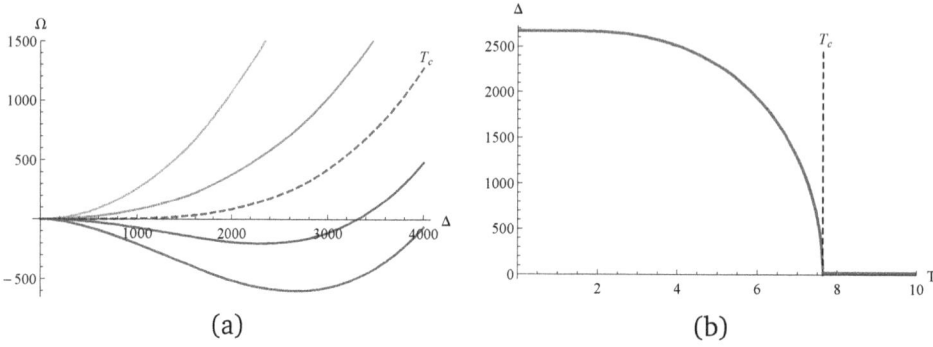

Figure 1.: (a) Effective potential for different values of the temperature. (b) Gap as a function of the temperature. The dashed line corresponds to the critical temperature.

3. Supersymmetric BCS

Let us now try to design an $\mathcal{N} = 1$ supersymmetric Lagrangian which incorporates these basic features. We are interested in a supersymmetric theory with a global baryonic $U(1)_B$ or $U(1)_R$ symmetry, which undergoes spontaneous symmetry breaking caused by fermion condensation triggered by quantum effects. Quartic fermion interactions arise by means of the following choice of Kähler potential:

$$K(\Phi, \Phi^\dagger) = \Phi^\dagger \Phi + g^2 (\Phi^\dagger \Phi)^2 \,. \tag{6}$$

We would like to construct a supersymmetric BCS theory with Dirac fermions, and in $\mathcal{N} = 1$ supersymmetric theories this requires at least two chiral superfields (a single chiral superfield describes a Weyl fermion). The simplest theory consists of two chiral superfields X and Y with the Kähler potential

$$K(X, Y, X^\dagger, Y^\dagger) = X^\dagger X + Y^\dagger Y + g^2 (X^\dagger X)^2 + g^2 (Y^\dagger Y)^2 \,. \tag{7}$$

The coupling g could in principle be different for the interaction terms involving X and Y superfields. One could also add, for example, a term $X^\dagger X Y^\dagger Y$ (used in [6]). However we shall consider the above simple choice which already illustrates the essential points.

3.1. Chemical potential for $U(1)_B$

We first consider the $\mathcal{N} = 1$ supersymmetric model defined in terms of two chiral superfields with Kähler potential (7) and superpotential:

$$W = mXY \,. \tag{8}$$

This gives masses to scalars and fermions. It will be shown that this model is *not* suitable to implement BCS mechanism in supersymmetric theories. The model will illustrate the typical problems that one has to deal with.

Once the Lagrangian corresponding to this model is written in terms of the component fields of the chiral multiplet, we need to introduce a chemical potential and consistency demands that this is coupled to a conserved non-anomalous U(1) current. The superfields X and Y carry opposite U(1) charge so the baryonic U(1)$_B$ current is non-anomalous. Turning on a chemical potential corresponds to turning on a background U(1)$_B$ gauge field component $A_0 = \mu$. This is done in the standard way, by replacing partial derivatives with covariant derivatives, $\partial_\nu \to D_\nu = \partial_\nu - iq\mu\delta_{\nu 0}$ (with no loss of generality one can set the X U(1) charge $q_X = 1$, as it can be absorbed into a redefinition of μ; in this way Y has charge $q_Y = -1$). In order to have a Lagrangian quadratic in fermion fields, one can make a Hubbard-Stratonovich transformation in the component Lagrangian by introducing two auxiliary fields Δ_x, Δ_y. As a result, the functional integral over fermions can be directly performed. Finally, we expand the scalar fields around their VEVs: $\phi = v + \varphi$, and retain only up to quadratic terms in the scalar fields (we assume real v). We find

$$
\mathcal{L}_S = (1 + 4g^2 v_x^2)\partial_\mu \varphi_x^* \partial^\mu \varphi_x + 4g^2 v_x^2 \left(\mu^2 - \frac{4g^2 m^2 v_y^2}{\left(1 + 4g^2 v_x^2\right)^3} \right)(\varphi_x^2 + \varphi_x^{*2})
$$

$$
+ \frac{4g^2 m^2 v_x v_y}{\left(1 + 4g^2 v_x^2\right)^2}(\varphi_x \varphi_y + \varphi_x^* \varphi_y + \varphi_x \varphi_y^* + \varphi_x^* \varphi_y^*) - \frac{m^2}{1 + 4g^2 v_x^2}|\varphi_y|^2
$$

$$
+ \left(\left(1 + 16g^2 v_x^2\right)\mu^2 - 4g^4|\Delta_x|^2 - \frac{4g^2 m^2 \left(-1 + 4g^2 v_x^2\right) v_y^2}{\left(1 + 4g^2 v_x^2\right)^3} \right)|\varphi_x|^2
$$

$$
+ i\mu(1 + 8g^2 v_x^2)(\varphi_x^* \partial_t \varphi_x - \varphi_x \partial_t \varphi_x^*) - 4i\mu g^2 v_x^2(\varphi_x^* \partial_t \varphi_x^* - \varphi_x \partial_t \varphi_x)
$$

$$
+ (x \leftrightarrow y, \mu \to -\mu),
$$

(9)

$$
\mathcal{L}_F = i(1 + 4g^2 v_x^2)(\psi_x^\dagger \bar{\sigma}^\mu \partial_\mu \psi_x) + \mu(1 + 8g^2 v_x^2)(\psi_x^\dagger \bar{\sigma}^0 \psi_x)
$$

$$
+ \left(\left(\frac{2mg^2 v_x v_y}{1 + 4g^2 v_x^2} + g^2\Delta_x \right)(\psi_x \psi_x) - \frac{1}{2}m\psi_x \psi_y + \text{h.c.} \right)
$$

(10)

$$
+ (x \leftrightarrow y, \mu \to -\mu),
$$

with classical potential

$$
V_{cl} = \frac{m^2 v_y^2}{1 + 4g^2 v_x^2} + (1 + 4g^2 v_x^2)(g^2|\Delta_x|^2 - \mu^2 v_x^2) + (x \leftrightarrow y).
$$

(11)

To have canonically normalized kinetic terms, one can redefine fields as

$$
\varphi \to \frac{\varphi}{\sqrt{1 + 4g^2 v^2}}, \qquad \psi \to \frac{\psi}{\sqrt{1 + 4g^2 v^2}}.
$$

(12)

Integrating over $\psi, \psi^\dagger, \varphi, \varphi^*$ leads to a one-loop potential, dependent on $g, v, \Delta, \mu,$ m. Since the model is not renormalizable (just like BCS), integrals will be regularized by a momentum cutoff, representing a "Debye" energy where new microscopic physics appears.

We proceed as follows. Calling O_S and O_F the resulting 4×4 scalar and fermion matrices for the quadratic terms in momentum space (see [1] for details), we shall write the determinants as:

$$\det O_S = \prod_{i=1}^{4} (\omega^2 - \omega_{S,i}^2), \qquad \det O_F = \prod_{i=1}^{4} (\omega^2 - \omega_{F,i}^2), \tag{13}$$

where

$$\omega_{S,i} = \omega_{S,i}(\mu, |\mathbf{p}|, g, m, v_x, v_y, \Delta_x, \Delta_y),$$
$$\omega_{F,i} = \omega_{F,i}(\mu, |\mathbf{p}|, g, m, v_x, v_y, \Delta_x, \Delta_y).$$

The eigenvalues for the frequencies have complicated expressions when v_x and v_y are non-vanishing. The strategy is to look for non-trivial minima at $v_x = v_y = 0$ with $\Delta_x, \Delta_y \neq 0$, assuming them to be real. Next, we shall check that the one-loop effective potential is locally stable in the v_x and v_y directions, a property that will be ensured by the presence of a mass term.

When $v_x = v_y = 0$, the scalar and fermion quadratic terms greatly simplify. At this point, we find the following eigenvalues for the frequency:

$$\omega_{S,1,2} = \sqrt{4g^4\Delta_x^2 + m^2 + p^2} \pm \mu,$$
$$\omega_{S,3,4} = \sqrt{4g^4\Delta_y^2 + m^2 + p^2} \pm \mu, \tag{14}$$

$$\omega_{F,1,2}^2 = 2g^4\Delta_x^2 + 2g^4\Delta_y^2 + \mu^2 + m^2 + p^2 \pm 2\mathcal{E}_+,$$
$$\omega_{F,3,4}^2 = 2g^4\Delta_x^2 + 2g^4\Delta_y^2 + \mu^2 + m^2 + p^2 \pm 2\mathcal{E}_-, \tag{15}$$

$$\mathcal{E}_\pm^2 = \mu^2\left(m^2 + p^2\right) + g^8\left(\Delta_x^2 - \Delta_y^2\right)^2 + g^4\left(m^2\left(\Delta_x + \Delta_y\right)^2 \pm 2\mu p\left(\Delta_x^2 - \Delta_y^2\right)\right). \tag{16}$$

For configurations with $\Delta_x = \Delta_y \equiv \Delta$, the fermion frequencies become

$$\omega_F = \sqrt{\left(\sqrt{p^2 + m^2 + 4g^4\Delta^2\frac{m^2}{\mu^2}} \pm \mu\right)^2 + 4g^4\Delta^2\left(1 - \frac{m^2}{\mu^2}\right)}. \tag{17}$$

On the other hand, for $\Delta_x = -\Delta_y \equiv \Delta$, we find

$$\omega_F = \sqrt{\left(\sqrt{p^2 + m^2} \pm \mu\right)^2 + 4g^4\Delta^2}. \tag{18}$$

This is the same dispersion relation as in the relativistic BCS system of section 2. This might suggest that the BCS mechanism can be implemented in a similar way. But the presence of charged scalars demands a careful approach. We first need to identify the Fermi surfaces. For $\Delta_x = \Delta_y = 0$, they lie on the region where $\omega_{F,2,4}$ vanish, i.e. at

$$\sqrt{p_F^2 + m^2} = \mu.$$

(19)

As in the standard relativistic BCS case, the existence of a Fermi surface would require $\mu > m$. However, in the present supersymmetric system we cannot set $\mu > m$ because the scalar contribution to the thermal partition function,

$$\frac{1}{\beta} \sum_i \int \frac{d^3p}{(2\pi)^3} \log\left(1 - e^{-\beta \omega_{S,i}}\right),$$

(20)

is ill-defined, because $\omega_{S,2,4}$ becomes negative below some value of the momentum. The system presents BE condensation, the occupation number of scalars with zero momentum goes to infinity as μ approaches m from below. This spoils the BCS mechanism.

One possible approach to elude this problem is to put the theory on $S^1 \times S^3$. Because scalars couple to curvature (see e.g. [7]), this will provide an extra mass for the scalar fields of the order of R^{-1} (R is the radius of the three-sphere). Taking $1/R > \Lambda$, the scalar contribution would be negligible, which may allow for regions in parameter space with Fermi surfaces, and without problems of BE condensation. However, this extra mass term is of the same order as the quantized fermion momentum values. Therefore it is not possible to separate the scalar mass scale from the Fermi surface, making BE condensation unavoidable.

3.2. A simple supersymmetric BCS model: Chemical potential for U(1)$_R$

Let us now consider an $\mathcal{N} = 1$ supersymmetric model with two chiral superfields X and Y, with Kähler potential given by (7) and superpotential $W = 0$. The Lagrangian has a U(1)$_R$ symmetry for arbitrary U(1)$_R$ charges of the X and Y superfields. It is convenient to consider the U(1)$_R$ symmetry under which scalars ϕ_x and ϕ_y are neutral, so that fermions ψ_x and ψ_y have charge -1. This choice avoids problems of BE condensation. Note that with this charge assignation the U(1)$_R$ symmetry is anomalous. However, this can be easily cured by adding to the theory free superfields with canonical Kähler potential with the required U(1)$_R$ charges to cancel the anomaly. For example, one may add Z_i ($i = 1, 2$), with R-charges $R(Z_i) = 2$, so that ψ_{Z_1}, ψ_{Z_2} have charges $+1$. The scalars in Z_i would then couple to the chemical potential and may undergo BE condensation. However, this sector is completely decoupled and therefore does not participate in the thermodynamics governing the X, Y sector.

In this case, the quadratic Lagrangian for the fluctuations (after expanding around expectation values) is given by

$$\mathcal{L}_S = \partial_\mu \varphi_x^* \partial^\mu \varphi_x + \partial_\mu \varphi_y^* \partial^\mu \varphi_y - \frac{4g^4 |\Delta_x|^2}{1 + 4g^2 v_x^2} |\varphi_x|^2 - \frac{4g^4 |\Delta_y|^2}{1 + 4g^2 v_y^2} |\varphi_y|^2, \tag{21}$$

$$\mathcal{L}_F = i(\psi_x^\dagger \bar{\sigma}^\mu \partial_\mu \psi_x) + i(\psi_y^\dagger \bar{\sigma}^\mu \partial_\mu \psi_y) - \mu(\psi_x^\dagger \bar{\sigma}^0 \psi_x) - \mu(\psi_y^\dagger \bar{\sigma}^0 \psi_y)$$

$$+ \left(\frac{g^2 \Delta_x}{1 + 4g^2 v_x^2} (\psi_x \psi_x) + \frac{g^2 \Delta_y}{1 + 4g^2 v_y^2} (\psi_y \psi_y) + h.c. \right), \tag{22}$$

where we have rescaled the fields to have canonical kinetic terms. The classical potential is given by

$$V_{cl} = g^2 \left(4g^2 v_x^2 + 1 \right) \left| \Delta_x \right|^2 + (x \leftrightarrow y). \tag{23}$$

The equations of motion for Δ_x, Δ_y give (setting the scalar fluctuations φ_x, φ_y to zero)

$$\Delta_x = \frac{\psi_x^\dagger \psi_x^\dagger}{(1 + 4g^2 v_x^2)^2}, \qquad \Delta_y = \frac{\psi_y^\dagger \psi_y^\dagger}{(1 + 4g^2 v_y^2)^2}. \tag{24}$$

Δ_x, Δ_y have both $U(1)_R$ charges equal to 2. Vacuum expectation values for them thus spontaneously break $U(1)_R$ and represent a measure of the fermion condensate.

By proceeding in a similar way as in the previous case, we now find the following frequencies for scalars and fermions:

$$\omega_{S,1,2}^2 = p^2 + \frac{4g^4 \Delta_x^2}{1 + 4g^2 v_x^2}, \qquad \omega_{S,3,4}^2 = p^2 + \frac{4g^4 \Delta_y^2}{1 + 4g^2 v_y^2}, \tag{25}$$

$$\omega_{F,1,2}^2 = (p \pm \mu)^2 + \frac{4g^4 \Delta_x^2}{\left(1 + 4g^2 v_x^2 \right)^2}, \qquad \omega_{F,3,4}^2 = (p \pm \mu)^2 + \frac{4g^4 \Delta_y^2}{\left(1 + 4g^2 v_y^2 \right)^2}. \tag{26}$$

Here we have chosen real Δ_x, Δ_y, as one-loop potential depends only on their moduli. We stress that these simple dispersion relations are a consequence of the extreme simplicity of this supersymmetric model; generic models (even with simple superpotentials) typically lead to very complicated eigenvalues for the frequencies.

In the present case, the dynamics of the X and Y fields are decoupled. It is clear that the same configuration that minimizes the one-loop potential in the x direction also minimizes the one-loop potential in the y direction. Therefore with no loss of generality we set $v_x = v_y \equiv v$ and $\Delta_x = \Delta_y \equiv \Delta$.

The complete one-loop thermodynamic potential is given by

$$\Omega = 2g^2 \left(1 + 4g^2 v^2\right) \Delta^2 + \frac{1}{\pi^2 \beta} \int\limits_0^\Lambda dp \, p^2 \left(2\log\left[\sinh\frac{\beta}{2}\omega_S\right]\right.$$

$$\left. - \log\left[\cosh\frac{\beta}{2}\omega_{F+}\right] - \log\left[\cosh\frac{\beta}{2}\omega_{F-}\right]\right). \tag{27}$$

When the vacuum lies at $\Delta \neq 0$, then $v = 0$ is a local minimum. When the vacuum lies at $\Delta = 0$, then there is a flat direction in v, because in this case the frequencies do not depend on v. This is confirmed by the evaluation of the one-loop potential.

The gap equation $\Delta = \Delta(T)$ is determined by the equation

$$\frac{\mathrm{d}\Omega}{\mathrm{d}\Delta} = 0. \tag{28}$$

This gives, when $v = 0$,

$$1 = \frac{g^2}{2\pi^2} \int\limits_0^\Lambda dp \, p^2 \left(\frac{\tanh\left(\frac{1}{2}\beta\sqrt{4g^4\Delta^2 + (p-\mu)^2}\right)}{\sqrt{4g^4\Delta^2 + (p-\mu)^2}}\right.$$

$$\left. + \frac{\tanh\left(\frac{1}{2}\beta\sqrt{4g^4\Delta^2 + (p+\mu)^2}\right)}{\sqrt{4g^4\Delta^2 + (p+\mu)^2}} - \frac{2\coth\left(\frac{1}{2}\beta\sqrt{4g^4\Delta^2 + p^2}\right)}{\sqrt{4g^4\Delta^2 + p^2}}\right). \tag{29}$$

This gap equation can be compared with the gap equation (5) of the relativistic BCS system of section 2. One difference is that now scalars and fermions have zero mass, since a mass term mXY would not be consistent with scalars being neutral under $U(1)_R$. The second and more fundamental difference is given by the scalar contribution – represented by the last term in the above equation – that we analyze in what follows.

An important consequence of supersymmetry is that the critical curve $\Delta(T)$ now depends logarithmically with the cutoff: for large p, the integral in (29) behaves as

$$\frac{g^2}{\pi^2} \int\limits_0^\Lambda dp \, \frac{\mu^2}{p} \sim \log \Lambda. \tag{30}$$

If the scalar contribution is removed, like in non-supersymmetric BCS, one has instead

$$\frac{g^2}{\pi^2} \int\limits_0^\Lambda dp \, p \sim \Lambda^2. \tag{31}$$

Obviously, a logarithmic dependence with the UV cutoff is a desirable feature, since the thermodynamics becomes much less sensitive to the underlying microscopic physics.

At the same time, the IR physics produced by the scalar sector has a striking effect: the superconducting transition becomes first-order, instead of second-order, as it would be in standard BCS. The IR physics of the scalar sector is important at the onset of the transition, where Δ is small. To see the nature of the transition, we need to compute $d\Delta/dT$. This can be obtained by differentiating the gap equation $d\Omega/d\Delta$ with respect to T. Writing the gap equation in the form $1 = f(\Delta^2, T)$, one has

$$\frac{d\Delta}{dT} = -\frac{\frac{\partial^2 \Omega}{\partial T \partial \Delta}}{\frac{\partial^2 \Omega}{\partial \Delta^2}} = -\frac{1}{2\Delta} \frac{\frac{\partial f}{\partial T}}{\frac{\partial f}{\partial(\Delta^2)}}. \tag{32}$$

In a second-order phase transition, $d\Delta/dT$ is singular at the critical temperature, where $\Delta = 0$. This is because $\partial f/\partial T$ and $\partial f/\partial(\Delta^2)$ are regular at $\Delta = 0$. While the scalar contribution to the one-loop potential is regular at $\Delta = 0$, its second derivative with respect to Δ^2 has a singularity near $\Delta = 0$ originating from the region near $p = 0$. We have

$$\frac{\partial f}{\partial(\Delta^2)} \approx \frac{8g^6 T}{\pi^2} \int_0^{} dp \, p^2 \frac{1}{(4\Delta^2 g^4 + p^2)^2} \approx \frac{g^4 T}{\pi} \frac{1}{\Delta}. \tag{33}$$

As a result, $d\Delta/dT$ is now finite at $\Delta = 0$. The superconducting phase transition is therefore first-order. This significant change coming from the $p = 0$ region would obviously not take place if the scalar field was massive. In such a case, the phase transition would still be second-order. But, as explained, in the present model it is not possible to add a mass term.

In an interval of temperature $T_{c1} < T < T_{c2}$ there are three branches – characteristic of first-order phase transitions – corresponding to three solutions of the gap equation: the trivial minimum at $\Delta = 0$, a maximum and another minimum at higher Δ. These are exhibited in figs. 2a and b, showing the one-loop potential at different temperatures and the corresponding gap as a function of the temperature. We see how the non-trivial maximum and minimum are created as the temperature is lowered below a certain critical value T_{c2}. At a temperature T_c, the non-trivial minimum becomes degenerate with the minimum at $\Delta = 0$. In the interval $T_{c1} < T < T_c$, the symmetric vacuum $\Delta = 0$ is metastable. And below T_{c1}, the symmetric vacuum $\Delta = 0$ becomes unstable and the only minimum of the potential is the SSB vacuum at $\Delta \neq 0$.

4. Conclusions

The salient aspects of this investigation are as follows. The main obstacle to implement BCS superconductivity in a neat way is, as expected, Bose-Einstein condensation. The

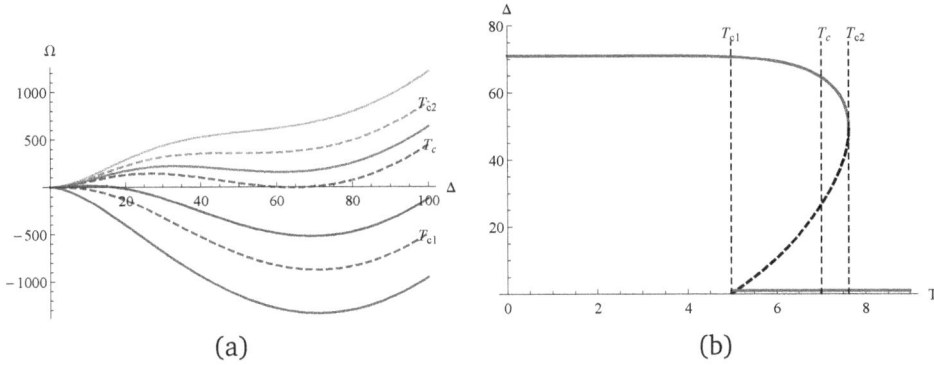

Figure 2.: (a) Effective potential for different values of the temperature. (b) Gap as a function of the temperature. The dashed lines corresponds to the critical temperatures T_c, T_{c1} and T_{c2}.

model of section 3.1 exhibits the typical problems that arise. Introducing chemical potential for a baryonic $U(1)_B$ symmetry leads to the emergence of Fermi surfaces, but inevitably couples scalar fields to the chemical potential as well, since scalar fields have the same baryon charge as fermions. Supersymmetry prevents the scalar mass scale from getting separated from the fermion mass scale.

To circumvent these problems, in section 3.2 we proposed a specific supersymmetric model which realizes BCS superconductivity. It is based on a Kähler potential with quartic terms in the superfields.

$$K = X^\dagger X + Y^\dagger Y + g^2 (X^\dagger X)^2 + g^2 (Y^\dagger Y)^2, \tag{34}$$

and $W = 0$. The chemical potential is introduced for a $U(1)_R$ symmetry under which the scalar components of the X and Y fields have vanishing charge. In this way Bose-Einstein condensation does not occur. This is presumably the simplest supersymmetric model for BCS superconductivity that one can construct, since it contains the minimum number of superfields necessary to have a Dirac fermion, i.e. two chiral superfields; and the superpotential is $W = 0$. We found that the system has a superconducting phase transition below some critical temperature, produced by a fermion condensate.

The equations determining the temperature dependence of the gap are very similar to the ones in BCS theory, with the main difference represented by the contribution coming from scalar fluctuations. One important effect of this contribution is a drastic reduction of the dependence on the UV cutoff, from quadratic to logarithmic. Another effect due to the scalar superpartner is changing the character of the phase transition from second to first-order. As explained, the origin of the change is the contribution of low momentum scalar modes at small Δ.

Acknowledgements

A.B. is supported by MECD FPU Grant No AP2009-3511. J.R. acknowledges support from project FPA 2010-20807.

References

[1] A. Barranco, J. G. Russo, "Supersymmetric BCS", *JHEP* **2012** (6 2012), 1.

[2] D. Bailin, A. Love, "Superfluidity and superconductivity in relativistic fermion systems", *Phys. Rep.* **107**.6 (1984), 325.

[3] D. Bertrand, "A Relativistic BCS Theory of Superconductivity: An Experimentally Motivated Study of Electric Fields in Superconductors" (2005), *(PhD Thesis)*.

[4] J. Terning, "Modern Supersymmetry: Dynamics and Duality", Oxford University Press (New York), 2006.

[5] J. Polchinski, "Effective Field Theory and the Fermi Surface" (1992), *(Lectures presented at TASI 1992)*, arXiv:9210046.

[6] W. Buchmüller, S. T. Love, "Chiral symmetry and supersymmetry in the Nambu-Jona-Lasinio model", *Nucl. Phys.* **B204**.2 (1982), 213.

[7] G. Festuccia, N. Seiberg, "Rigid supersymmetric theories in curved superspace", *JHEP* **2011** (6 2011), 1.

The many surprises of maximal supergravity

Andrea Borghese[1,a], **Adolfo Guarino**[2,b] **and Diederik Roest**[3,a]

[a]Centrum voor Theoretische Natuurkunde, Rijksuniversiteit Groningen, Nijenborgh 4, 9747 AG Groningen, Netherlands

[b]Albert Einstein Center for Fundamental Physics, Institute for Theoretical Physics, Universität Bern, Sidlerstrasse 5, 3012 Bern, Switzerland

E-mail: [1]a.borghese@rug.nl, [2]guarino@itp.unibe.ch, [3]d.roest@rug.nl

Abstract

We describe recent developments regarding gauged $\mathcal{N} = 8$ supergravity in $D = 4$. Using the embedding tensor formulation we show how to classify all the extrema of this theory with a G_2 residual gauge symmetry. Our classification contains all the vacua of the recently discovered [1] family of SO(8) gauged maximal supergravities.

1. Introduction

Maximal supergravities in four dimensions have been discovered at the end of the seventies and, after more than thirty years, still have surprises in store.

The ungauged version [2] can be obtained by torus reduction of eleven dimensional supergravity. Many distinctive features already show up at this level. In particular the field content is completely fixed and consists of the $\mathcal{N} = 8$ supergravity multiplet alone without any possibility of coupling matter multiplets. The theory contains 70 real scalar fields which can be viewed as coordinates of an $E_7/SU(8)$ coset manifold. During the last decade ungauged maximal supergravity has received some attentions because of its special ultraviolet behaviour. Four-graviton scattering amplitudes are finite up to four loops [3]. This could be a hint of the fact that the theory is perturbatively finite providing the first and only instance of an ultraviolet finite, point-like theory of quantum gravity.

In a gauged supergravity some vectors in the spectrum are used to gauge a subgroup of the duality group E_7. The gauging procedure is the only known consistent deformation of maximal supergravity. The main advantage of carrying out this procedure is the introduction of a potential which enriches the dynamics in the scalar sector.

The first example of this class of maximal supergravities is the SO(8) gauged theory [4] in which all the 28 physical vectors in the spectrum are used to gauge an SO(8) subgroup of E$_7$. This gauged theory can be obtained by sphere reduction of eleven dimensional supergravity and has been intensively studied from many perspectives. Its vacuum structure has been analysed restricting to truncated sectors [5] and, in recent years, a scan of all vacua having cosmological constant within a certain range of values has been performed using numerical techniques [6]. With the advent of the gauge/gravity duality the SO(8) gauged maximal supergravity has been used to construct models of holographic superconductivity [7–10]. Furthermore the theory should be dual to the ABJM three dimensional CFT [11].

Some other gauged maximal supergravities have been constructed with a variety of gauge groups [12, 13]. Theories with a compact gauge group usually displays AdS vacua while, whenever a non-compact group is gauged, it is possible to find dS critical points though perturbatively unstable[1]. The quest for a stable dS vacuum in maximal supergravity has been, so far, inconclusive[2]. Either this difficulty could be overcome by spotting the right gauging or it could be a hint of a fundamental obstruction against the realisation of dS vacua in maximal supergravity.

Gauged supergravity plays also a role in the field of flux compactifications. As described above, some gaugings could be obtained by compactifications of higher dimensional theories. Other gaugings do not have a higher dimensional origin but show interesiting features. it is worth understanding which higher dimensional ingredients (compactification manifold, fluxes or extended sources such as branes or orientifold planes) need to be used to obtain a certain gauged supergravity.

Due to the web of open questions and interesting applications it is worthwhile investigating the vacua of gauged maximal supergravity. This could give the possibility of answering some of those questions or provide models for holography. Our contribution takes a step in this direction.

Despite the fact that only a handful of gaugings have been worked out explicitly, in the last decade a gauging independent formulation of the theory [16, 17] has been constructed. This formulation makes use of an object called embedding tensor, originally introduced in the context of three dimensional supergravity [18, 19], which allows to keep implicit the gauge group and still write down a Lagrangian as we will explain in section 2. The embedding tensor formulation also allows to use group theory as a powerful tool to understand the structure of gauged supergravity. An example of this is the new family of SO(8) gauged supergravity discovered in [1]. Up to 2012,

---

[1]dS/Minkowski critical points are perturbatively unstable whenever, expanding around those points, some scalar excitation has a negative squared mass. For AdS critical point, a scalar field is unstable if its mass squared is below the Breitenlohner-Freedman (BF) bound [14]. In our conventions, instability means $m^2 L^2 = 3m^2/|V| \leq -9/4$.

[2]However see [15] for the last interesting development in this direction.

the SO(8) gauged theory was believed to be unique. Using group theory it is instead easy to show that there is a one parameter family of such theories with a different vacuum structure. This finding will be described in section 3. Another interesting development which gives the possibility of understanding more clearly the vacuum structure of gauged supergravity is what we will call the "going to the origin" (GTTO) approach [20, 21]. We will describe this approach in section 4 and show how it allows to classify exhaustively vacua preserving a given amount of symmetry. In section 5 we will go through the example of critical points with a residual G_2 symmetry [22] and explain the relation between our findings and the new family of SO(8) maximal supergravity found in [1].

2. Embedding tensor formulation

The field content of $\mathcal{N} = 8$ supergravity in $D = 4$ consists of a graviton, 8 gravitini, 28 vectors, 56 spin-1/2 fields and 70 scalar fields. In this section we briefly review the most recent construction of the gauged theory with the aim of underlining some fundamental concepts. We refer to [16, 17] for a more detailed and explicit treatment.

The 70 scalar fields are associated to isometries of the scalar manifold $E_7/SU(8)$. This means they are in a one to one correspondence with the 70 non compact generators of E_7. We denote the adjoint of E_7 with indices $\alpha, \beta \in \mathbf{133}$. The generators t_α split in 63 compact and 70 non-compact ones.

The theory construction relies on the E_7 duality group. The 28 physical vectors, which we call electric, are supplemented with their magnetic counterpart and together they sit in the fundamental representation of the duality group. We write them as $A_\mu{}^{\mathcal{M}}$ with $\mathcal{M} \in \mathbf{56}$ of E_7. Some of these vectors are used to gauge a subgroup $G_g \subset E_7$. The structure of derivatives on a generic field is hence modified according to

$$\mathcal{D}_\mu = \partial_\mu - A_\mu{}^{\mathcal{M}} \Theta_{\mathcal{M}}{}^\alpha t_\alpha . \tag{1}$$

The object denoted with Θ is called embedding tensor. It selects the linear combination of vectors which is used to gauge a particular isometry of the scalar manifold. We can look at it as a set of charges which transform non trivially under the action of the duality group. In particular $\Theta \in \mathbf{56} \times \mathbf{133}$ of E_7.

In order for the gauging procedure to be consistent we need to impose some constraints on the embedding tensor: a set of linear constraints (LC) and a set of quadratic ones (QC). The LC are necessary to preserve invariance of the action under supersymmetry. They amount to ask the embedding tensor to belong to a single representation of E_7

$$\Theta \in \mathbf{912} \subset \mathbf{56} \times \mathbf{133} . \tag{2}$$

The QC are instead necessary for the consistency of the gauging. In fact it is of fundamental importance that the charges of a gauge theory be invariant under the action of the gauge group. In our case the gauge group is a subgroup of the duality group. Thus we have to impose explicitly that, while transforming under E_7, the embedding tensor be invariant under the action of $G_g \subset E_7$. This could be written in the following form

$$\Theta_M{}^\alpha \Theta_N{}^\beta \Omega^{MN} = 0,$$ (3)

where Ω is antisymmetric and invariant under the action of E_7. If we define an element of the gauge algebra with $\mathfrak{X}_M = \Theta_M{}^\alpha t_\alpha$ and the generalised structure constants with $\mathfrak{X}_{MN}{}^P = \Theta_M{}^\alpha [t_\alpha]_N{}^P$, equation (3) implies

$$[\mathfrak{X}_M, \mathfrak{X}_N] = -\mathfrak{X}_{MN}{}^P \mathfrak{X}_P,$$ (4)

where with $[t_\alpha]_N{}^P$ we mean that the generator is in the E_7 fundamental representation. Equation (4) implies the closure of the gauge algebra.

Once embedding tensor components are chosen satisfying the LC and the QC, the gauge group is fixed and a consistent G_g gauged supergravity is obtained. Nevertheless it is not compulsory to specify the embedding tensor components. The gauged Lagrangian in [16, 17] is perfectly consistent and invariant under local supersymmetry transformations even if the embedding tensor is left unspecified but satisfies the constraints (2, 3). In this sense the construction is gauging independent and gives the universal Lagrangian of four dimensional maximal supergravity.

We will give now some more detail regarding the scalar sector of the theory. Scalar fields will be denoted with ϕ. A central role in the theory construction is played by the scalar manifold coset representative. We will denote it with $\mathcal{V}(\phi)_M{}^{\underline{M}}$. This object is an element of the E_7 group. We can act on it with a global E_7 transformation from the left (on the index M) and with local $SU(8) \subset E_7$ transformations from the right (on the index $\underline{M}$). it is used to couple gauge field strength (which have global indices) to fermionic degrees of freedom (which have $SU(8)$ local indices being $SU(8)$ the R-symmetry group). From the coset representative $\mathcal{V}$ it is possible to define a metric on the scalar manifold $\mathcal{M} = \mathcal{V}\mathcal{V}^T$ and hence the scalar potential in its gauging independent form

$$V(\phi, \Theta) = \frac{1}{672} \left(\mathfrak{X}_{MN}{}^R \mathfrak{X}_{PQ}{}^S \mathcal{M}^{MP} \mathcal{M}^{NQ} \mathcal{M}_{RS} + 7 \mathfrak{X}_{MN}{}^Q \mathfrak{X}_{PQ}{}^N \mathcal{M}^{MP} \right).$$ (5)

it is worth pointing out some features of equation (5). First of all it is invariant under duality transformations in the following sense: for every transformation U acting on the scalars through the coset representative $U_M{}^N \mathcal{V}(\phi)_N{}^{\underline{N}} = \mathcal{V}(\phi')_M{}^{\underline{N}}$, there is a compensating transformation on the embedding tensor $U\Theta = \Theta'$ such that $V(\phi', \Theta) = V(\phi, \Theta')$. Furthermore the scalar potential has in general a complicated non linear dependence on the scalar fields while it is homogeneous and quadratic in the embedding tensor components through the generalised structure constants. These considerations will prove useful in the next sections.

3. The new family of SO(8) gaugings

We will now specify the gauge group to be SO(8) and show how the family of SO(8) gauged maximal supergravities comes out from group theoretical considerations. We follow the argument of [1] in the derivation.

As explained in the previous section, the embedding tensor components should be singlets under the gauge group. Thus we need to consider the branching of the **912** representation of E_7 duality group under SO(8)

$$912 = 2 \times \left(1 + 35_v + 35_s + 35_c + 350\right), \tag{6}$$

where the **35** representations are related to the vector, spinor and conjugate spinor representation of SO(8). In there we find two singlets and the parameter interpolating between these singlets span the family of SO(8) gauged theories. Despite the simplicity of this argument it took thirty years to acknowledge that in fact there is not just a single theory but rather a one parameter family of different theories.

At this point we introduce a bit of notation. We consider the SL(8) symplectic frame [16] in which the 56 vectors can be split in 28 electric ones transforming in the **28** of SL(8) and 28 magnetic transforms in the contravariant representation, the **28'**. We denote vectors as $A_\mu{}^{\mathcal{M}} = \{A_\mu{}^{[AB]}, A_{\mu\,[AB]}\}$ with $A, B \in \mathbf{8}$ of SL(8). The SO(8) group is trivially embedded in SL(8) and the two singlets in (6), coupled to electric and magnetic vectors respectively, can be written as

$$\Theta_{\mathcal{M}}{}^{\alpha} = \{\Theta_{AB}{}^{C}{}_{D}, \Theta^{ABC}{}_{D}\} = \{\cos\omega\,\delta^{C}_{[A}\,\delta_{B]D}, \sin\omega\,\delta^{[A}_{D}\,\delta^{B]C}\}. \tag{7}$$

As already explained, the ω parameter interpolates between electric and magnetic components of the embedding tensor.

For different values of ω we are in front of different maximal supergravity theories with SO(8) gauge group. In order to show this and make contact with the next sections we consider a truncation of these theories: the restriction to the G_2-invariant sector [23]. As the name suggests, the truncation contains all fields in the spectrum which are invariant under a subgroup of the duality group, namely the G_2 term in

$$E_7 \supset SL(2) \times G_2, \tag{8}$$

while other fields are set to zero. The truncation is $\mathcal{N} = 1$ supersymmetric, in the bosonic sector there are no vectors and only one complex field z is retained. The scalar potential (5) reduces to

$$V = e^{\mathcal{K}}\left[\mathcal{K}^{z\bar{z}}\left(\mathcal{D}_z W\right)\left(\mathcal{D}_{\bar{z}}\overline{W}\right) - 3W\overline{W}\right], \tag{9}$$

with

$$\mathcal{K} = -7\ln\left[-1 + \frac{1}{1+z} + \frac{1}{1+\bar{z}}\right], \quad \mathcal{W} = \frac{\sqrt{2}}{(1+z)^7}\left[(1+7z^4)e^{i\omega} + (7z^3 + z^7)e^{-i\omega}\right].$$

(10)

The following plots show the form of the scalar potential in terms of two real fields $\{\phi_1, \phi_2\}$ with $z = -(\phi_1 + i\phi_2)$ for two distinct values $\omega = 0, \pi/8$.

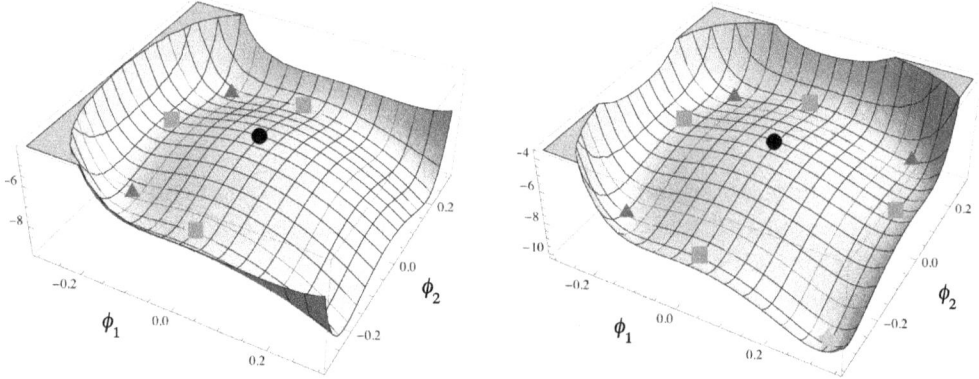

Figure 1.: Scalar potential of the new SO(8) gauged maximal supergravities in the G_2 invariant sector. On the left $\omega = 0$ while on the right $\omega = \pi/8$.

The left plot in Figure 1 contains six critical points divided in three different groups. At the origin there is an $\mathcal{N} = 8$ supersymmetric critical point preserving the full gauge symmetry group SO(8) marked with a black circle. The three points marked with red squares are non supersymmetric critical points in which the gauge group is broken from SO(8) to SO(7). Finally the blue triangles are $\mathcal{N} = 1$ supersymmetric critical points preserving a residual G_2 symmetry.

Three new critical points appear in the right plot of Figure 1. Two of them fall in the same groups which were present in the $\omega = 0$ case. The last one, marked with an orange star, preserves again a G_2 symmetry but is non supersymmetric.

These plots show that for different values of ω the potential (9) displays a different vacuum structure. In particular, the appearence of a new type of critical point (the non supersymmetric G_2) demonstrates that the two theories at $\omega = 0$ and $\pi/8$ are fundamentally different.

Even though the vacuum positions and the cosmological constant values change with ω, the normalised squared masses $m^2 L^2$ of the two scalar excitations remain the same for critical points preserving the same amount of supersymmetry and residual gauge symmetry. In particular, for the maximally supersymmetric critical point we have $m^2_{1,2}L^2 = -2$. For the $\mathcal{N} = 1$, G_2 symmetric ones we have $m^2_{1,2}L^2 = 4 \pm \sqrt{6}$. For the non supersymmetric SO(7) invariant critical points we have $m^2_1 L^2 = 6$ and $m^2_2 L^2 = -6/5$. Finally for the $\mathcal{N} = 1$, G_2 symmetric one we have $m^2_{1,2}L^2 = 6$.

146

There is a motivation for the fact that moving in the ω space we find different theories. To explain that we need to go back to the definition of duality group. Such a group is defined as the invariance group of the set of equations of motions (EOM) plus Bianchi identities (BI) of a supergravity theory. Acting with the duality group we do change the Lagrangian of the theory but we do not change the set of EOM plus BI thus leaving unchanged physical observables (such that critical points). Hence it is clear that an ω transformation must sit outside E_7. Indeed they belong to the complement of E_7 inside $Sp(56) \supset E_7$. The symplectic group $Sp(56)$ is the electro-magnetic (E-M) duality group rotating among each others electric and magnetic vectors [1].

4. Going to the origin

In this section we will describe an alternative approach for the study of vacuum structure in supergravity [20]. The GTTO approach relies on the special form of extended supergravity scalar manifolds. In particular for (half)-maximal supergravities these are always homogeneous coset manifolds G/H. For such spaces there is no preferred point due to the transitive action of G on the elements of the manifold. A way of rephrasing this sentence is stating that every point of the coset can be transformed to the origin[3] of the coset.

In the origin of moduli space we loose covariance w.r.t. the full E_7 group and we are left with $SU(8)$ covariant objects. Every tensor in a given E_7 representation branches in different $SU(8)$ representations. In this language there are two embedding tensor components corresponding to the two terms in the branching $912 = 36 + 420 + \text{c.c.}$

$$\Theta_{\mathcal{M}}{}^{\alpha} \;\rightarrow\; A^{IJ} \in \mathbf{36}, \quad A_I{}^{JKL} \in \mathbf{420} \quad \text{with } I,J,K,L \in \mathbf{8} \text{ of } SU(8). \tag{11}$$

The scalar potential (5) can be written in the elegant form

$$V = -\frac{3}{8} A^{IJ} A_{IJ} + \frac{1}{24} A_I{}^{JKL} A^I{}_{JKL}. \tag{12}$$

Notice that the nonlinear dependence on scalar fields has disappeared leaving us with a simple expression quadratic in the embedding tensor components.

Usually we are interested in finding critical points of (5) for a fixed gauging (thus for fixed embedding tensor components). Suppose for a given gauging Θ_0 we find a critical point $\phi = \phi_0$ for which $\partial_\phi V(\phi_0, \Theta_0) = 0$. At the price of modifying the form of the embedding tensor we can bring the critical point to the origin of moduli space using a field independent E_7 transformation. This means that, leaving the embedding tensor free, every critical point could be brought to the origin.

---

[3]By definition, the origin of the coset space in a given parametrisation is identified with the point $\phi = 0$.

This is exactly the philosophy of the GTTO approach. We can choose embedding tensor components which are free to take any possible configuration invariant under a particular subgroup $G_r \subset E_7$ and compatible with the LC and QC. At the origin of moduli space we compute the first derivative of the scalar potential. Being at the origin, the EOM are just a set of equations quadratic in the embedding tensor components. We solve these equations finding the set of structure constants which are compatible with having a critical point at the origin $\partial_\phi V(0, \Theta_0') = 0$.

In this way we are sure to find all critical points of gauged maximal supergravity which have as residual gauge symmetry the group G_r. Clearly the critical point we have found at the origin with the particular set of embedding tensor components Θ_0' could be related to another critical point at the position $\phi = \phi_0$ but with a simpler set of structure constants Θ_0, but it is difficult to tackle this correspondence. In other words the drawback of this procedure is that, in principle, we have lost contact with the full gauge group G_g.

5. All G_2 invariant critical points

We are now ready to show how, using the GTTO approach, it is possible to classify all critical points preserving a certain amount of gauge symmetry. We fix the residual symmetry to be at least G_2. This means the embedding tensor components need to be constructed using G_2 singlets.

In order to parametrise G_2 invariant tensors, we split indices according to

$$I = (1, m), \tag{13}$$

where $m, n, \ldots$ is the fundamental of G_2. The latter is also the fundamental of SO(7) when embedded in SO(8) in the standard way. G_2 is defined as the subgroup of SO(7) that leaves invariant a particular 3-form φ_{mnp} and its dual 4-form in seven dimensions. Decomposing the **36** and **420** of SU(8) into G_2, one finds two and three singlets, respectively. We will parametrise these with the following Ansatz:

$$A^{IJ} \rightarrow A^{11} = \alpha_1, \quad A^{mn} = \alpha_2 \delta^{mn}, \tag{14}$$
$$A_I{}^{JKL} \rightarrow A_1{}^{mnp} = \beta_1 \varphi^{mnp}, \quad A_m{}^{1np} = \beta_2 \varphi_m{}^{np}, \quad A_m{}^{npq} = \beta_3 (*\varphi)_m{}^{npq}.$$

Plugging the most general ansatz with five complex parameters into the QC and the EOM one gets a number of quadratic constraints on these parameters. As explained in more detail in [20], these are amenable to an exhaustive analysis by means of algebraic geometry techniques, in particular prime ideal decomposition, and the corresponding code SINGULAR [24]. In this way we find the four branches of solutions listed below, all corresponding to Anti-de Sitter space-times. In all cases we will omit an overall scaling of the solutions and use SU(8) to set the phases of α_1 and α_2 equal. They are either $\mathcal{N} = 0, 1$ or 8:

- The first branch is $\mathcal{N} = 8$ and reads

$$\alpha = \left(e^{i\Theta}, e^{i\Theta}\right), \quad \beta = (0,0,0). \tag{15}$$

All solutions are SO(8) invariant and preserve $\mathcal{N} = 8$. They correspond to the origin of the standard SO(8) gauging and its one-parameter generalisation. As also noted in [1], the mass spectrum is equal for the entire branch and given by

$$m^2 L^2 = -2 \, (\times 70), \tag{16}$$

- The second branch is $\mathcal{N} = 1$ and is given by

$$\alpha = \left(-2e^{-5i\Theta}, \sqrt{6}e^{-5i\Theta}\right), \quad \beta = \left(0, \sqrt{2/3}e^{-i\Theta}, e^{3i\Theta}\right). \tag{17}$$

For all values of these parameters, the invariance group is $G_r = G_2$ and the mass spectrum reads

$$m^2 L^2 = (4 \pm \sqrt{6}) \, (\times 1), \quad 0 \, (\times 14), \quad -\tfrac{1}{6}(11 \pm \sqrt{6}) \, (\times 27). \tag{18}$$

This coincides with the G_2 invariant mass spectra of the standard SO(8) theory. The latter corresponds to a particular value of Θ. Other values include the one-parameter generalisation of [1] and possibly more.

- The third branch is $\mathcal{N} = 0$ and reads

$$\alpha = \left(3e^{-3i\Theta}, 3e^{-3i\Theta}\right), \quad \beta = \left(-e^{i\Theta}, e^{i\Theta}, \mp e^{i\Theta}\right). \tag{19}$$

The stability group in this case is $G_r = SO(7)_\pm$. The mass spectrum is independent of the parameter and reads

$$m^2 L^2 = 6 \, (\times 1), \quad 0 \, (\times 7), \quad -\tfrac{6}{5} \, (\times 35), \quad -\tfrac{12}{5} \, (\times 27). \tag{20}$$

The lowest of the eigenvalues violates the BF bound and hence this branch is perturbatively unstable. The spectrum coincides with the $SO(7)_\pm$ invariant mass spectra of the standard SO(8) theory, as was later explained group theoretically in [25]. Again, the latter corresponds to a single point in a one-dimensional parameter space of non supersymmetric $SO(7)_\pm$ invariant critical points.

- The last branch is $\mathcal{N} = 0$ as well and is given by

$$\alpha = \left(\sqrt{3}e^{-3i\Theta}, -e^{-3i\Theta}\right), \quad \beta = \left(e^{i\Theta}, \tfrac{1}{3}\sqrt{3}e^{i\Theta}\right). \tag{21}$$

The invariance group is $G_r = G_2$ and the mass spectrum reads

$$m^2 L^2 = 6 \, (\times 2), \quad 0 \, (\times 14), \quad -1 \, (\times 54). \tag{22}$$

In this case all eigenvalues satisfy the BF bound, and hence this family of critical points is non-supersymmetric and nevertheless perturbatively stable. Previously known examples of stability without supersymmetry were isolated points with smaller symmetry groups [9, 26].

it is easy to compare the results in this section with the ones in the last section. We see that all critical points appearing in Fig.1 with the related symmetries and mass spectra fall in one of the listed branches. There is nontheless one fundamental difference. The ω parameter of section 3 describes theories having the same SO(8) gauge group. Within this one parameter family of theories we can follow the evolution of every critical point of the potential (9). In this section we are following a different evolution. For every branch the parameter Θ describes the evolution of a critical point with fixed residual gauge symmetry and supersymmetry along the set of different theories which are compatible with it.

We will better clarify this point by analysing more specifically a particular branch, e.g. the SO(7)$_+$ one. In order to show the physical changes when traversing the Θ space we have calculated the eigenvalues of the Cartan-Killing metric, from which the full gauge group (and not only the invariance group of the critical point) can be derived. Again, this is outlined in [20] and we employ the mapping given in [26]. For the SO(7)$_+$ critical points a set of 21 eigenvalues is always negative, corresponding to the preserved part of the gauge group. The remaining 7 are either all negative, zero or positive, as a function of Θ, leading to the following gauge groups:

$$\Theta \in \Big[0, \arccos \sqrt{\tfrac{1}{6}(3 + \sqrt{5})}\Big) : \qquad\qquad G_g = \text{SO}(7,1),$$
$$= \arccos \sqrt{\tfrac{1}{6}(3 + \sqrt{5})} : \qquad\qquad G_g = \text{ISO}(7,1),$$
$$\in \Big(\arccos \sqrt{\tfrac{1}{6}(3 + \sqrt{5})}, \tfrac{\pi}{4}\Big] : \qquad\qquad G_g = \text{SO}(8). \qquad (23)$$

The gauge group therefore changes from compact to non-compact and viceversa, while passing trought an İnönü-Wigner contracted point.

In this sense the GTTO approach contains the classical approach as the Θ parameter describes more than just the SO(8) gauged theory critical points.

6. Conclusion

In this contribution we have briefly described some of the latest developments in the field of gauged maximal supergravity in four dimensions. After introducing the embedding tensor formulation of the theory, we have explained how the new one parameter family of SO(8) gauged supergravity has been discovered opening up a new landscape of theories. Finally we have shown how the GTTO approach could be used as a powerful tool for the study of vacuum structure.

Many questions are still waiting for an answer. First of all we need to understand whether this additional parameter is present for other gaugings, how big is the parameter space of new theories and what is the periodicity of this space. This would be of crucial importance providing new explicit examples to study with new interesting properties for many purposes. Related to this, it is still an open problem to identify the higher dimensional origin of the ω phase for the SO(8) gauged theory. Finally we need to understand how far we can go using the GTTO approach. In principle this approach allows to classify vacua with a generic residual symmetry. Unfortunately, the smaller the residual symmetry group the more expensive in terms of computational power is the task.

Acknowledgements

AB would like to thank the organizers of the Barcelona Postgrad Encounters on Fundamental Physics for the beautiful conference and the warm hospitality. The research of AB and DR is supported by a VIDI grant from the Netherlands Organisation for Scientific Research (NWO). The work of AG is supported by the Swiss National Science Foundation.

References

[1]   G. Dall'Agata, G. Inverso, M. Trigiante, "Evidence for a family of SO(8) gauged supergravity theories", *Phys. Rev. Lett.* **109** (2012), 201301.

[2]   E. Cremmer, B. Julia, "The SO(8) Supergravity", *Nucl. Phys.* **B159** (1979), 141.

[3]   Z. Bern et al., "Amplitudes and Ultraviolet Behavior of N = 8 Supergravity", *Fortsch.Phys.* **59** (2011), 561.

[4]   B. de Wit, H. Nicolai, "N=8 Supergravity", *Nucl. Phys.* **B208** (1982), 323.

[5]   N. Warner, "Some new extrema of the scalar potential of gauged N=8 supergravity", *Phys. Lett.* **B128** (1983), 169.

[6]   T. Fischbacher, "The Encyclopedic Reference of Critical Points for SO(8)-Gauged N=8 Supergravity. Part 1: Cosmological Constants in the Range -$\Lambda/g^2 \in [6 : 14.7)$" (2011), arXiv:1109.1424.

[7]   A. Donos, J. P. Gauntlett, "Superfluid black branes in $AdS_4 \times S^7$", *JHEP* **1106** (2011), 053.

[8]   N. Bobev et al., "Supergravity Instabilities of Non-Supersymmetric Quantum Critical Points", *Class. Quant. Grav.* **27** (2010), 235013.

[9]   T. Fischbacher, K. Pilch, N. P. Warner, "New Supersymmetric and Stable, Non-Supersymmetric Phases in Supergravity and Holographic Field Theory" (2010), arXiv:1010.4910.

[10] N. Bobev et al., "Minimal Holographic Superconductors from Maximal Supergravity", *JHEP* **1203** (2012), 064.

[11] O. Aharony et al., "N=6 superconformal Chern-Simons-matter theories, M2-branes and their gravity duals", *JHEP* **0810** (2008), 091.

[12] C. Hull, N. Warner, "The potential of the gauged N=8 supergravity theories", *Nucl. Phys.* **B253** (1985), 675.

[13] C. Hull, "New gauged N=8, D = 4 supergravities", *Class. Quant. Grav.* **20** (2003), 5407.

[14] P. Breitenlohner, D. Z. Freedman, "Positive Energy in anti-De Sitter Backgrounds and Gauged Extended Supergravity", *Phys. Lett.* **B115** (1982), 197.

[15] G. Dall'Agata, G. Inverso, "de Sitter vacua in N = 8 supergravity and slow-roll conditions", *Phys. Lett.* **B718** (2013), 1132.

[16] B. de Wit, H. Samtleben, M. Trigiante, "Magnetic charges in local field theory", *JHEP* **0509** (2005), 016.

[17] B. de Wit, H. Samtleben, M. Trigiante, "The Maximal D=4 supergravities", *JHEP* **0706** (2007), 049.

[18] H. Nicolai, H. Samtleben, "Maximal gauged supergravity in three-dimensions", *Phys. Rev. Lett.* **86** (2001), 1686.

[19] H. Nicolai, H. Samtleben, "Compact and noncompact gauged maximal supergravities in three-dimensions", *JHEP* **0104** (2001), 022.

[20] G. Dibitetto, A. Guarino, D. Roest, "Charting the landscape of N=4 flux compactifications", *JHEP* **1103** (2011), 137.

[21] G. Dall'Agata, G. Inverso, "On the Vacua of N = 8 Gauged Supergravity in 4 Dimensions", *Nucl. Phys.* **B859** (2012), 70.

[22] A. Borghese, A. Guarino, D. Roest, "All G_2 invariant critical points of maximal supergravity", *JHEP* **1212** (2012), 108.

[23] A. Borghese et al., "The SU(3)-invariant sector of new maximal supergravity" (2012), arXiv:1211.5335.

[24] W. Decker et al., "Singular 3-1-2 — A computer algebra system for polynomial computations" ().

[25] H. Kodama, M. Nozawa, "Classification and stability of vacua in maximal gauged supergravity", *JHEP* **1301** (2013), 045.

[26] G. Dibitetto, A. Guarino, D. Roest, "Exceptional Flux Compactifications", *JHEP* **1205** (2012), 056.

HvLif solutions in ungauged supergravity

P. Bueno[1,a], **W. Chemissany**[2,b], **P. Meesen**[3,c], **T. Ortín**[4,a] and **C. S. Shahbazi**[5,a,d]

[a]Instituto de Física Teórica, Universidad Autónoma de Madrid,
  C/ Nicolás Cabrera 13-15, Cantoblanco, 28049 Madrid, Spain

[b]Department of Physics and Astronomy, University of Waterloo, Waterloo,
  N2L 3G1 Ontario, Canada

[c]HEP Theory Group, Departamento de Física, Universidad de Oviedo,
  Avda. Calvo Sotelo s/n, 33007 Oviedo, Spain

[d]Stanford Institute for Theoretical Physics and Department of Physics,
  Stanford University, Stanford, CA 94305-4060, USA

E-mail: [1]pab.bueno@estudiante.uam.es, [2]chemissany.wissam@gmail.com,
[3]meessenpatrick@uniovi.es, [4]tomas.ortin@csic.es, [5]carlos.shabazi@uam.es

Abstract

In this note we describe a procedure to construct, from known black-hole and black-brane solutions of any ungauged supergravity theory, non-trivial gravitational solutions whose "near-horizon" and "near-singularity" limits are Lifshitz-like spacetimes with dynamical critical exponent z and a "hyperscaling violation" exponent θ and Lifshitz radius ℓ that depends on the physical parameters of the original black-hole solution. Since the new Lifshitz-like solutions can be constructed from any black-hole solution of any ungauged supergravity, many of them can be easily embedded in String Theory.

1. Introduction

Gauge/gravity duality has found new and interesting applications in the study of strongly coupled condensed matter systems [1–3]. In this context one has to work with the metrics that are dual to scale-covariant field theories which are not conformally invariant. These theories are characterized by a dynamical critical exponent $z \neq 1$ and a *hyperscaling violation* exponent $\theta \neq 0$ [4–8]. The values $z = 1$ and $\theta = 0$ correspond to conformally-invariant theories dual to the AdS metric. For other values of z and

θ, in terms of dimensionless coordinates t, r, x^i, the $(d + 2)$-dimensional spacetime metric can be cast in the form

$$ds^2_{d+2} = \ell^2 r^{-2(d-\theta)/d} \left[r^{-2(z-1)} dt^2 - dr^2 - dx^i dx^i \right] , \qquad (1)$$

where d is the number of spatial dimensions on which the dual theory lives ($i = 1, \ldots, d$) and the parameter ℓ, with dimensions of length, is the *Lifshitz radius*. We will refer, henceforth, to these metrics as *hyperscaling-violating Lifshitz (hvLif) metrics*.

hvLif geometries (1) with the particular hyperscaling violation exponent $\theta = d - 1$ are intimately connected with compressible states with hidden Fermi surfaces as well as with logarithmic violations of the area law of entanglement entropy (see for instance [6–9]). hvLif solutions have also been of interest for their connection with string theory and supergravity. We refer the reader to *e.g.* [7, 10–12].

It is known since the advent of the AdS/CFT duality that considering temperature in the gauge theory corresponds to putting a black hole in the bulk of the gravity side [13, 14]. When the gauge theory is conformal, the corresponding black hole has AdS asymptotics in the boundary. Similarly, since Lifshitz field theory with hyperscaling violation has anisotropic scale covariance, the black holes describing the geometry dual to its finite temperature generalization must have a metric of the form (1) as asymptotic geometry.

Asymptotically Lifshitz black holes with $\theta = 0$ and $z \neq 1$ have been extensively studied over the last few years. Analytic and numerical solutions to gravitational theories with simple types of matter have been constructed. String theoretic black hole solutions having Lifshitz asymptotics with $\theta = 0$ with a general dynamical exponent have been numerically computed (see [15, 16] and references therein). Analytical black hole solutions which could be related to string and supergravity theories are still missing, though[1].

So far, hvLif metrics (1) with $\theta \neq 0$ have only been found in solutions to Einstein-Maxwell-dilaton-type effective actions of the form

$$S = \frac{1}{16\pi G_N} \int \sqrt{|g|} \left\{ R + \tfrac{1}{2} \partial_\mu \phi \partial^\mu \phi - Z(\phi) F^{\mu\nu} F_{\mu\nu} - 2\Lambda - V(\phi) \right\} . \qquad (2)$$

Simple analytic black hole solutions have been constructed for this model for specific choices of $V(\phi)$ and $Z(\phi)$ in [6–9, 18]; analytical hvLif solutions to a model with 2 gauge fields and an exponential scalar potential, are presented and analysed in ref. [19]. Finding embeddings of these models and solutions in gauged supergravities and, eventually, in string theory would be most interesting, in particular for asymptotically hvLif $\theta = d - 1$ black holes.

---

[1] The analytic black hole solutions in [17] suffer from naked singularities.

In this work we report progress in this direction. In particular, we are going to show how to construct systematically solutions of ungauged supergravity whose metrics are, or approach in certain limits, hvLif metrics with certain values of z and θ. Our construction makes use of the FGK formalism originally developed to study static, spherically symmetric, asymptotically flat, black hole solutions of 4-dimensional ungauged supergravity theories [20], and we start by reviewing this formalism in Section 2. We will then generalize the FGK formalism to metrics which are not spherically symmetric. The main result is that there are (at least) two cases in which the equations of motion of the metric function and scalar fields are identical to those of the spherically symmetric one. Thus, one can use the solutions of the standard black hole case and construct solutions with entirely different spacetime metrics.

In section 3 we study the behaviour of the new solutions in the neighborhood of the values of the radial coordinate corresponding, in the original solution, to the inner and outer horizons, spatial infinity and the curvature singularity. We will find hvLif metrics in some of these limits. In Section 5 we conclude.

2. The generalized FGK formalism

Following Ref. [20] we consider the four-dimensional action

$$I = \int \left\{ R + \mathcal{G}_{ij}(\phi)\partial_\mu\phi^i\partial^\mu\phi^j + 2\mathfrak{Im}\,\mathcal{N}_{\Lambda\Sigma}F^\Lambda{}_{\mu\nu}F^{\Sigma\,\mu\nu} - 2\mathfrak{Re}\,\mathcal{N}_{\Lambda\Sigma}F^\Lambda{}_{\mu\nu}\star F^{\Sigma\,\mu\nu} \right\}, \quad (3)$$

where $\mathcal{N}_{\Lambda\Sigma}$ is the complex scalar-dependent (*period*) matrix, and (only in (3)) $\int \equiv \int d^4x\sqrt{|g|}$. The bosonic sector of any ungauged supergravity theory in 4 dimensions can be put in this form. The number of scalars labeled by $i, j, \cdots$ and of vector fields labeled by $\Lambda, \Sigma, \cdots$, the scalar metric $\mathcal{G}_{ij}$ and the period matrix $\mathcal{N}_{\Lambda\Sigma}$ depend on the particular theory under consideration.

Since we want to obtain static solutions, we consider the metric

$$ds^2 = e^{2U}dt^2 - e^{-2U}\gamma_{\underline{mn}}dx^{\underline{m}}dx^{\underline{n}}, \quad (4)$$

where $\gamma_{\underline{mn}}$ is a 3-dimensional (*transverse*) Riemannian metric to be specified later. Using Eq. (4) and the assumption of staticity of all the fields, we perform a dimensional reduction over time in the equations of motion that follow from the above general action. We obtain a set of reduced equations of motion, two of which can be obtained from the three-dimensional effective action[2]

$$I = \int d^3x\sqrt{|\gamma|}\left\{ R + \mathcal{G}_{AB}\partial_{\underline{m}}\tilde{\phi}^A\partial^{\underline{m}}\tilde{\phi}^B \right\}, \quad (5)$$

---

[2]See Ref. [20] for more details on this reduction.

whereas the remaining one reads

$$\partial_{[m}\psi^\Lambda \partial_{n]}\chi_\Lambda = 0 \,. \tag{6}$$

All the tensor quantities refer to the 3-dimensional metric $\gamma_{\underline{mn}}$, and we have defined the metric $\mathcal{G}_{AB}$

$$\mathcal{G}_{AB} \equiv \begin{pmatrix} 2 \\ & \mathcal{G}_{ij} \\ & & 4e^{-2U}\mathcal{M}_{MN} \end{pmatrix} , \tag{7}$$

in the *extended* manifold of coordinates $\tilde{\phi}^A = \left(U, \phi^i, \psi^\Lambda, \chi_\Lambda \right)$, where

$$(\mathcal{M}_{MN}) \equiv \begin{pmatrix} (\mathfrak{I}+\mathfrak{R}\mathfrak{I}^{-1}\mathfrak{R})_{\Lambda\Sigma} & -(\mathfrak{R}\mathfrak{I}^{-1})_\Lambda{}^\Sigma \\ -(\mathfrak{I}^{-1}\mathfrak{R})^\Lambda{}_\Sigma & (\mathfrak{I}^{-1})^{\Lambda\Sigma} \end{pmatrix}, \quad \mathfrak{R}_{\Lambda\Sigma} \equiv \mathfrak{Re}\,\mathcal{N}_{\Lambda\Sigma}, \quad \mathfrak{I}_{\Lambda\Sigma} \equiv \mathfrak{Im}\,\mathcal{N}_{\Lambda\Sigma}, \tag{8}$$

and ψ^Λ, χ_Λ are the electrostatic and magnetostatic potentials, respectively.

If we now decide to consider spherically-symmetric transverse metrics only, as it is appropriate to describe single, static black holes, we can choose, as in Ref. [20]

$$\gamma_{\underline{mn}}\,\mathrm{d}x^{\underline{m}}\mathrm{d}x^{\underline{n}} = \frac{\mathrm{d}\tau^2}{W_{-1}^4} + \frac{\mathrm{d}\Omega_{-1}^2}{W_{-1}^2}, \tag{9}$$

where W_{-1} is a function of the (inverse) radial coordinate τ to be determined and

$$\mathrm{d}\Omega_{-1}^2 \equiv \mathrm{d}\theta^2 + \sin^2\theta\,\mathrm{d}\phi^2 \tag{10}$$

is the metric of the round 2-sphere of unit radius. With this choice, Eq. (6) is automatically solved, and the equation of $W_{-1}(\tau)$ can be integrated completely, giving

$$W_{-1}(\tau) = \frac{\sinh\left(r_0\tau\right)}{r_0}, \tag{11}$$

where the integration constant r_0 is the *non-extremality parameter*: when r_0 vanishes, the metric describes extremal black holes (if the solution satisfies the necessary regularity conditions).

The electrostatic and magnetostatic potentials $\psi^\Lambda, \chi_\Lambda$ only appear through their τ-derivatives. The associated conserved quantities are the magnetic and electric charges p^Λ, q_Λ and can be used to eliminate completely the potentials. The remaining equations of motion can be derived from the effective action

$$I_{\mathrm{eff}}[U, \phi^i] = \int \mathrm{d}\tau \left\{ (U')^2 + \tfrac{1}{2}\mathcal{G}_{ij}\phi^{i\prime}\phi^{j\prime} - e^{2U}V_{\mathrm{bh}} \right\}, \tag{12}$$

to which we have to add the Hamiltonian constraint

$$(U')^2 + \tfrac{1}{2}\mathcal{G}_{ij}\phi^{i\prime}\phi^{j\prime} + e^{2U}V_{\text{bh}} = r_0^2, \tag{13}$$

which expresses the conservation of the Hamiltonian (due to absence of explicit τ-dependence of the Lagrangian) but with a particular value of the integration constant (r_0^2). The primes indicate differentiation with respect to τ and the so-called *black-hole potential* V_{bh} is given by[3]

$$-V_{\text{bh}}(\phi, \mathcal{Q}) \equiv -\tfrac{1}{2}\mathcal{Q}^M \mathcal{Q}^N \mathcal{M}_{MN}, \qquad (\mathcal{Q}^M) \equiv \begin{pmatrix} p^\Lambda \\ q_\Lambda \end{pmatrix}. \tag{14}$$

A fair number of solutions of this system for different theories of $\mathcal{N} = 2, d = 4$ supergravity coupled to vector supermultiplets are known (see *e.g.* Ref. [21, 22]). They describe single, charged, static, spherically-symmetric, asymptotically-flat, non-extremal black holes which generalize the Reissner-Nordström solution and have two horizons that coincide when the non-extremality parameter r_0 vanishes. The metric covers the exterior of the outer (event) horizon when the (*inverse*) radial coordinate[4] τ takes values in the interval $(-\infty, 0)$, whose limits are, respectively, the event horizon and spatial infinity. The interior of the inner (Cauchy) horizon corresponds to the interval $(\tau_s, +\infty)$, whose limits are, respectively, the singularity and the inner horizon.

We may also be interested in spacetime metrics which are not spherically symmetric, in which case we have to use a different transverse metric. In principle, these metrics are not appropriate to describe isolated, static black holes but here we are ultimately interested in Lifshitz metrics with a transverse metric invariant under the 2-dimensional Euclidean group. Thus, we can take, for instance, the following simple generalization of the spherically-symmetric transverse metric Eq. (9):

$$\gamma_{\underline{mn}} dx^{\underline{m}} dx^{\underline{n}} = \frac{d\tau^2}{W_\kappa^4} + \frac{d\Omega_\kappa^2}{W_\kappa^2}, \tag{15}$$

where W_κ is a function of τ and $d\Omega_\kappa^2$ is the metric of the 2-dimensional symmetric space of curvature κ and unit radius:

$$d\Omega_{-1}^2 \equiv d\theta^2 + \sin^2\theta\, d\phi^2; \quad d\Omega_{+1}^2 \equiv d\theta^2 + \sinh^2\theta\, d\phi^2; \quad d\Omega_0^2 \equiv d\theta^2 + d\phi^2. \tag{16}$$

In these three cases the equation for $W_\kappa(\tau)$ can be integrated and the results are

$$W_{-1} = \frac{\sinh r_0\tau}{r_0}; \quad W_1 = \frac{\cosh r_0\tau}{r_0}; \quad W_0^\pm = a e^{\mp r_0\tau}; \tag{17}$$

---

[3]As in Ref. [21], we adopt the sign of the black-hole potential opposite to most of the literature on black-hole attractors, conforming instead to the conventions of Lagrangian mechanics.

[4]Observe that τ has dimensions of inverse length, since r_0 has, conventionally, dimensions of length.

where a is a real arbitrary constant with dimensions of inverse length.

It turns out that if we follow now for the $\kappa = 0, +1$ cases the procedure described above for the $\kappa = -1$ case we arrive to exactly the same system of equations and, therefore, to the same effective action Eq. (12). It follows that all the solutions for $\left(U, \phi^i \right)$ obtained in the spherically-symmetric case $\kappa = -1$ are also solutions for the $\kappa = 0, +1$ cases as well. In other words: every solution of the Hamiltonian constraint Eq. (13) and the system of equations obtained from Eq. (12) provides us with four different solutions of the original theory, by simply using the four different transverse metrics.

Since, as mentioned above, there exists a number of solutions of those equations that describe single, static, asymptotically-flat non-extremal black holes when we take $\kappa = -1$ [21–23], we can simply take those solutions and study them setting $\kappa = 0$ or $+1$ in the transverse metric. Observe that one integration constant has been fixed to normalize the metric at spatial infinity, something we may not need to do in the $\kappa = 0, +1$ cases, but the normalization could be changed at any moment, if necessary.

In what follows we are going to study the asymptotic behaviour of generic solutions (U, ϕ^i), normalized to describe single, static, asymptotically-flat non-extremal black holes for $\kappa = -1$ when we take the transverse metric with $\kappa = 0$.[5]

3. Solutions with Lifshitz-like asymptotics

Since we are going to use the metric functions e^{-2U} corresponding to charged, spherically-symmetric, asymptotically-flat, non-extremal black-hole solutions, we start by reviewing their asymptotic behaviors at the outer (+) and inner (−) horizons[6] (placed, respectively, at $\tau = -\infty$ and $\tau = +\infty$) and at spatial infinity $\tau = 0$.

- The standard normalization of these asymptotically-flat black holes requires that

$$\lim_{\tau \to 0^-} e^{-2U} = 1. \tag{18}$$

- When τ approaches the two horizons, $\tau \to \mp\infty$, the metric function behaves as

$$e^{-2U} \sim \frac{S_\pm}{4\pi r_0^2} e^{\mp 2r_0 \tau}, \tag{19}$$

where S_+ (resp. S_-) is the entropy of the outer (resp. inner) horizon, which is assumed to be non-vanishing (which is equivalent to require regularity of the

---

[5]We leave the remaining case ($\kappa = +1$) of the *triality* for a future publication.
[6]Uncharged, static black holes only have an outer horizon. The discussion of the behaviour of the metric function in the interior of the inner horizon does not apply to them.

black-hole solution). If we use the spherically-symmetric transverse metric the spacetime metric approaches in these limits a product of a Rindler metric and a 2-sphere of area $4S_\pm$. Studying the Rindler metric by conventional methods one finds that the temperatures of the horizons $T_\pm$ obey the Smarr-like relation [24]

$$r_0 = 2S_\pm T_\pm . \tag{20}$$

- e^{-2U} vanishes for some value of $\tau_s \in (0, +\infty)$ at which the physical singularity of the black-hole spacetime lies. We may also want to study the behaviour of e^{-2U} near this value of τ but we do not know of any general result on this respect. We will have to study each particular case separately.

To find new solutions, we are going to plug black-hole metric functions in the general static metric Eq. (4) with the transverse metric Eq. (15) with $\kappa = 0$ (see Eqs. (16) and (17)). It is convenient to set $a = 1/r_0$ so no new length scale is introduced in the metric, which takes two possible forms:

$$ds^2_{(\pm)} = e^{2U}dt^2 - e^{-2U}\left[e^{\pm 4r_0\tau}r_0^4\,d\tau^2 + e^{\pm 2r_0\tau}r_0^2\left(d\theta^2 + d\phi^2\right)\right]. \tag{21}$$

3.1. Asymptotic behaviour of $ds^2_{(-)}$

Using the general properties of the metric function e^{-2U} described above it is easy to see that in the limit $\tau \to -\infty$ this metric behaves as

$$ds^2_{(-)} \sim \frac{4\pi r_0^2}{S_+}e^{2r_0\tau}dt^2 - \frac{S_+}{4\pi r_0^2}e^{-2r_0\tau}\left[e^{-4r_0\tau}r_0^4\,d\tau^2 + e^{-2r_0\tau}r_0^2\left(d\theta^2 + d\phi^2\right)\right]. \tag{22}$$

The change of coordinates: $r \equiv e^{-r_0\tau}$, $\tilde{t} \equiv \frac{4\pi r_0^2}{S_+}t/r_0$, $x^1 \equiv \theta$, $x^2 \equiv \phi$ brings the metric to the form

$$ds^2_{(-)} \sim \frac{S_+}{4\pi}r^4\left[r^{-6}\,d\tilde{t}^2 - dr^2 - dx^i dx^i\right], \tag{23}$$

which is a hvLif metric of the form Eq. (1) with $z = 4$, $\theta = 6$ and radius $\ell \sim r_0$ up to dimensionless factors (functions of the quotient S_+/r_0^2); this asymptotic hvLif space lies in the class of Ricci flat hvLif spaces.

The metric $ds^2_{(-)}$ is regular at $\tau = 0$. Spatial infinity is not there because the radial distance between points with $\tau = 0$ and points with $\tau < 0$ is finite and not infinite, as in the black-hole case. For τ equal to a certain τ_s, $e^{-2U} = 0$ and the metric will be singular, as in the black-hole case. Finally, in the $\tau \to +\infty$ limit the metric is the product of Rindler spacetime times $\mathbb{R}^2$, which can be understood as a flat event horizon with the same temperature as that of the inner horizon of the associated black-hole solution.

3.2. Asymptotic behaviour of $ds^2_{(+)}$

The analysis is completely analogous to the previous case: in the limit $\tau \to -\infty$ we find a flat event horizon whose temperature is that of the outer horizon of the associated black hole, there is a singularity at $\tau = \tau_s$ and a hyperscaling Lifshitz metric in the $\tau \to +\infty$ limit. The Lifshitz radius is, once again, $\ell = r_0$.

4. Examples

4.1. The Schwarzschild black hole

This is the only uncharged, static, spherically-symmetric, black-hole solution of the class of theories we are considering and has only one horizon (the event horizon) at (conventionally) $\tau \to -\infty$ in these coordinates, which only cover the exterior. The metric function for the Schwarzschild black hole in these coordinates is $e^{-2U} = e^{2M\tau}$, whereas the spacetime metric $ds^2_{(-)}$ takes the explicit form

$$ds^2_{(-)} = e^{-2M\tau}dt^2 - e^{-2M\tau}M^4 d\tau^2 - M^2 \left(d\theta^2 + d\phi^2\right) . \tag{24}$$

This can be rewritten as

$$ds^2_{(-)} \overset{[e^{-M\tau}M \equiv r]}{=} r^2 d(t/M)^2 - dr^2 - M^2 \left(d\theta^2 + d\phi^2\right) , \tag{25}$$

which is the product of a 2-dimensional Rindler spacetime ($\mathcal{R}i^2$) with $\mathbb{R}^2$. The temperature of the flat horizon would be that of the Schwarzschild black hole $T \sim M^{-1}$. Observe that this is not just the asymptotic behaviour of the metric: the metric is everywhere identically $\mathcal{R}i^2 \times \mathbb{R}^2$. As is well-known, this metric is just a wedge of the Minkowski spacetime which can be recovered by analytical extension of this metric.

Observe that the above metric makes sense for $\tau \in (-\infty, +\infty)$ or $r \in (0, +\infty)$ since as discussed above, there is not spatial infinity at $\tau = 0$.

The metric $ds^2_{(+)}$ is in this case

$$ds^2_{(+)} = e^{-2M\tau}dt^2 - e^{6M\tau}M^4 d\tau^2 - e^{4M\tau}M^2 \left(d\theta^2 + d\phi^2\right) , \tag{26}$$

or

$$ds^2_{(+)} \overset{[e^{M\tau} \equiv r]}{=} r^{-2} d(t/M)^2 - r^4 M^2 dr^2 - r^4 M^2 \left(d\theta^2 + d\phi^2\right) \tag{27}$$
$$= M^2 r^4 \left\{ r^{-6} dt^2 - dr^2 - d\theta^2 - d\phi^2 \right\} ,$$

which is the $z = 4$, $\theta = 6$, $\ell \sim M$ hvLif metric everywhere in the spacetime, and not just asymptotically. Yet again, this metric is defined for all values of τ or for all $r \in (0, +\infty)$.

4.2. The Reissner-Nordström black hole

The embedding of the Reissner-Nordström black hole in pure $\mathcal{N} = 2, d = 4$ supergravity (the supersymmetrization of the Einstein-Maxwell theory) has a metric function which, in the τ coordinates, reads [21]

$$e^{-2U} = \left[\cosh r_0 \tau - \frac{M}{r_0} \sinh r_0 \tau \right]^2, \qquad r_0^2 \equiv M^2 - |\mathcal{Z}|^2, \qquad (28)$$

where $\mathcal{Z} = \frac{1}{2}p - iq$ is the central charge of pure $\mathcal{N} = 2, d = 4$ supergravity in the chosen conventions.

It is evident that the asymptotic behaviour of the metrics $ds^2_{(\pm)}$ fits in the general case discussed above. Having the explicit form of the metric, we can also study the behaviour of the spacetime metric near the singularity at τ_s at which $e^{-2U}(\tau_s) = 0$. It is, however, easier to do it in the coordinates in which the metric function has the standard form [7]

$$e^{-2U} = \frac{r^2}{(r - r_+)(r - r_-)}, \qquad r_\pm = M \pm r_0, \qquad (29)$$

$$ds^2_{(\pm)} = \frac{(r - r_+)(r - r_-)}{r^2} dt^2 - \frac{r_0^4 r^2}{(r - r_\pm)(r - r_\mp)^5} dr^2 - \frac{r_0^2 r^2}{(r - r_\mp)^2}(d\theta^2 + d\phi^2). \quad (30)$$

According to the general discussion, we should find the singularity in the extension of the metric beyond $\tau = 0$ to positive values of τ. This corresponds in these coordinates to values of r "beyond $r = +\infty$". Thus, we define the coordinate $u \equiv 1/r$ which overlaps with r for $u > 0$ and extends the metric for $u \leq 0$. In these coordinates the metric takes the form

$$ds^2_{(\pm)} = (1 - r_+ u)(1 - r_- u) dt^2 - \frac{r_0^4}{(1 - r_\pm u)(1 - r_\mp u)^5} dr^2 - \frac{r_0^2}{(1 - r_\mp u)^2}(d\theta^2 + d\phi^2),$$
$$(31)$$

and, in the $u \to -\infty$ limit it approaches the metric

$$ds^2_{(\pm)} = r_+ r_- u^2 dt^2 - \frac{r_0^4}{r_\pm r_\mp^5 u^6} du^2 - \frac{r_0^2}{r_\mp^2 u^2}(d\theta^2 + d\phi^2), \qquad (32)$$

which can be put in the hvLif form with $z = 3$, $\theta = 4$ with the coordinate change $r' \equiv 1/u$ using the rescaled coordinates $\tilde{t} \equiv r_\pm t/r_0^2$, $\rho \equiv r'/r_\mp$, $x^1 \equiv \sqrt{r_+ r_-}/r_0 \theta$, $x^2 \equiv \sqrt{r_+ r_-}/r_0 \phi$.

---

[7]The coordinate transformation that relates these two forms of the metric function is
$$r = -\left[\cosh r_0 \tau - \frac{M}{r_0} \sinh r_0 \tau \right] \left[\frac{\sinh r_0 \tau}{r_0} \right]^{-1}.$$

Observe that the two consecutive coordinate changes $r = 1/u$, $u = 1/r'$ mean that we can get the same result taking the limit of the metric when r approaches $r = 0$ (which corresponds to the value $\tau = \tau_s$) "from the left". In fact, the same result is obtained if we take the near-singularity limit from the right.

Summarizing, the interior of the inner horizon region $r < r_-$ has, therefore, two boundaries, at $r = r_-$ and at $r = 0$. When the metric approaches $r = r_-$ from the left, the metric $ds^2_{(+)}$ approaches a hvLif metric with $z = 4$ and $\theta = 6$ and the metric $ds^2_{(-)}$ approaches $\mathcal{R}i^2 \times \mathbb{R}^2$, as we have seen before. When r approaches $r = 0$ the metric approaches a hvLif metric with $z = 3$, $\theta = 4$.

The fact that a hvLif metric can describe the near-singularity limit of a metric that has been obtained as a deformation of a regular black-hole metric is very suggestive. Observe that the deformed metric Eq. (30) differs from the standard Reissner-Nordström metric in factors of $(r - r_\pm)$, which are irrelevant in the $r \to 0$ limit, and in the 2-dimensional metric $d\Omega^2_\kappa$ which has $\kappa = -1$ for the standard, spherically symmetric Reissner-Nordström black hole. In [25] it is shown that there is a limit of the Reissner-Nordström black hole in which $d\Omega^2_{-1}$ approaches $d\Omega^2_0$. The near-singularity limit of this Reissner-Nordström black hole turns out to be described by a hvLif metric with $z = 3$, $\theta = 4$.

5. Conclusions

In this paper we have shown how to construct solutions of $\mathcal{N} = 2$, $d = 4$ supergravity presenting a hvLif behavior in the near-horizon and near-singularity limits by *deforming* static, spherically-symmetric, asymptotically flat black hole solutions of the same theory, without solving any equation[8].

Since some hvLif are holographically related to some well-known QFTs, it would be interesting to find holographically related QFTs that could describe those classical curvature singularities, at least in some regime. Finding quantum systems with the right values of z and θ may be difficult, or impossible, though. More work is needed to see if this possibility can be realized.

Acknowledgements

The authors would like to thank E. Ó Colgáin and S. Kachru for stimulating discussions. WC and CSS would like to thank the Stanford Institute for Theoretical Physics for its hospitality; likewise, PM would like to thank the Instituto de Física Teórica its

---

[8]The D-dimensional generalization of the procedure described here, as well as many other examples can be found in [25].

hospitality. This work has been supported in part by the Spanish Ministry of Science and Education grant FPA2009-07692, the Principáu d'Asturies grant IB09-069, the Comunidad de Madrid grant HEPHACOS S2009ESP-1473, and the Spanish Consolider-Ingenio 2010 program CPAN CSD2007-00042. The work of PM has been supported by the Ramón y Cajal fellowship RYC-2009-05014. The work of PB and CSS has been supported by the JAE-predoc grants JAEPre 2011 00452 and JAEPre 2010 00613. TO wishes to thank M.M. Fernández for her constant support.

References

[1]   S. A. Hartnoll, "Lectures on holographic methods for condensed matter physics", *Class. Quant. Grav.* **26** (2009), 224002.

[2]   C. P. Herzog, "Lectures on Holographic Superfluidity and Superconductivity", *J. Phys.* **A42** (2009), 343001.

[3]   J. McGreevy, "Holographic duality with a view toward many-body physics", *Adv. High. Energy Phys.* **2010** (2010), 723105.

[4]   C. Charmousis et al., "Effective Holographic Theories for low-temperature condensed matter systems", *JHEP* **1011** (2010), 151.

[5]   B. Gouteraux, E. Kiritsis, "Generalized Holographic Quantum Criticality at Finite Density", *JHEP* **1112** (2011), 036.

[6]   L. Huijse, S. Sachdev, B. Swingle, "Hidden Fermi surfaces in compressible states of gauge-gravity duality", *Phys. Rev.* **B85** (2012), 035121.

[7]   X. Dong et al., "Aspects of holography for theories with hyperscaling violation", *JHEP* **1206** (2012), 041.

[8]   E. Shaghoulian, "Holographic Entanglement Entropy and Fermi Surfaces", *JHEP* **1205** (2012), 065.

[9]   N. Ogawa, T. Takayanagi, T. Ugajin, "Holographic Fermi Surfaces and Entanglement Entropy", *JHEP* **1201** (2012), 125.

[10]   E. Perlmutter, "Hyperscaling violation from supergravity", *JHEP* **1206** (2012), 165.

[11]   K. Narayan, "On Lifshitz scaling and hyperscaling violation in string theory", *Phys. Rev.* **D85** (2012), 106006.

[12]   M. Ammon, M. Kaminski, A. Karch, "Hyperscaling-Violation on Probe D-Branes", *JHEP* **1211** (2012), 028.

[13]   E. Witten, "Anti-de Sitter space and holography", *Adv. Theor. Math. Phys.* **2** (1998), 253.

[14]   E. Witten, "Anti-de Sitter space, thermal phase transition, and confinement in gauge theories", *Adv. Theor. Math. Phys.* **2** (1998), 505.

[15]   I. Amado, A. F. Faedo, "Lifshitz black holes in string theory", *JHEP* **1107** (2011), 004.

[16]   L. Barclay et al., "Lifshitz black holes in IIA supergravity", *JHEP* **1205** (2012), 122.

[17]   W. Chemissany, J. Hartong, "From D3-Branes to Lifshitz Space-Times", *Class. Quant. Grav.* **28** (2011), 195011.

[18]   N. Iizuka et al., "Holographic Fermi and Non-Fermi Liquids with Transitions in Dilaton Gravity", *JHEP* **1201** (2012), 094.

[19]   M. Alishahiha, E. O Colgain, H. Yavartanoo, "Charged Black Branes with Hyper-scaling Violating Factor" (2012), arXiv:1209.3946.

[20]   S. Ferrara, G. W. Gibbons, R. Kallosh, "Black holes and critical points in moduli space", *Nucl. Phys.* **B500** (1997), 75.

[21]   P. Galli et al., "Non-extremal black holes of N=2, d=4 supergravity", *JHEP* **1107** (2011), 041.

[22]   T. Mohaupt, O. Vaughan, "The Hesse potential, the c-map and black hole solutions", *JHEP* **1207** (2012), 163.

[23]   P. Meessen et al., "H-FGK formalism for black-hole solutions of N=2, d=4 and d=5 supergravity", *Phys. Lett.* **B709** (2012), 260.

[24]   G. W. Gibbons, R. Kallosh, B. Kol, "Moduli, scalar charges, and the first law of black hole thermodynamics", *Phys. Rev. Lett.* **77** (1996), 4992.

[25]   P. Bueno et al., "Lifshitz-like Solutions with Hyperscaling Violation in Ungauged Supergravity" (2012), arXiv:1209.4047.

Static, spherically symmetric, charged black holes in four-dimensional $\mathcal{N} = 2$ supergravity

Pietro Galli[1] and Gabriela Onandía[2]

Departament de Física Teòrica and IFIC, Universitat de València, C/ Dr. Moliner, 50, 46100 Burjassot (València), Spain

E-mail: [1]pietro.galli@ific.uv.es, [2]gabriela.onandia@uv.es

Abstract

Besides general relativity, charged black holes appear as solutions of the equations of motion in theories where gravity is coupled to Maxwell fields and massless scalars. To this case it belongs $\mathcal{N} = 2$, $D = 4$ supergravity and in these notes we aim to review its basis and some last recent developments in the study of static, spherically symmetric, black-hole solutions.

1. Introduction

Nowadays the best candidate of a theory of quantum gravity is thought to be string theory. Its low energy limit is supergravity, a gauge theory of both the spacetime symmetries (diffeomorphisms and Lorentz transformations) and supersymmetry. In supergravity, black-hole solutions for the spacetime metric arise quite naturally and in the last two decades lots of efforts have been put in the attempt to obtain their complete, final classification. However, while for supersymmetric solutions it can be said that a satisfactory understanding has been achieved, for non-supersymmetric black holes still some pieces of the puzzle have to be placed. The recent introduction of the *H-FGK formalism* [1] seems to simplify this issue and its resolution may not be too far.

This short review is structured as follows: in the next two sections one can find a general discussion about static black holes in general relativity and Einstein-Maxwell-scalar theories and the basis of the *FGK formalism* [2]. Section 4 reviews black holes in $\mathcal{N} = 2$, $D = 4$ supergravity pointing out the differences among the possible solutions and their first-order description. Sections 5 and 6 sketch the main ideas of the new H-FGK formulation of the theory and conclude these notes mentioning some recent results.

2. Black hole basics

In theoretical physics, black holes appeared originally as solutions of the Einstein equations of motion. The first exact black-hole solution was obtained by Schwarzschild [3] in 1915 only one month after the publication of the Einstein theory of gravity. It describes the geometry of the spacetime curved by a spherically symmetric non-rotating mass without any charge. Few months later Reissner solved the Einstein equations for a charged point mass [4] and short after Nordström extended this solution to a spherically symmetric charged body [5]. In 1965 Newman found a metric describing charged rotating black holes [6, 7] that merges the Kerr solution (achieved in 1963 [8]) for spinning point mass and the Reissner-Nordström one. The Kerr-Newman black-hole solution is characterized by the value of only three parameters: the mass, the total charge and the angular momentum. This is the statement of the *no-hair theorem* that holds true for Maxwell-Einstein theory in four dimensions and according to which all the information concerning the constituent matter, the birth process and objects at infinity do not affect the externally observable features of a black hole. This important theorem finds its expression in the macroscopic black hole thermodynamics first studied by Bekenstein and Hawking in the seventies. They managed to find a close resemblance between the laws of black hole mechanics and the laws of thermodynamics ending up proving that the entropy is proportional to the event horizon area while the temperature to the surface gravity [9–11].

In this review we are going to focus on static, spherically symmetric, charged black holes. In general relativity they are obtained by solving the Einstein-Maxwell equations:

$$R^{\mu\nu} = 8\pi \left(\mathcal{F}^{\mu\gamma}\mathcal{F}^{\nu}{}_{\gamma} + \tfrac{1}{4}G^{\mu\nu}\mathcal{F}_{\gamma\delta}\mathcal{F}^{\gamma\delta} \right), \tag{1}$$

with the ansatz for the metric $G_{\mu\nu}$:

$$ds^2 = e^{2U(\tau)}dt^2 - e^{-2U(\tau)} \left(r_0^4 \sinh^{-4}(r_0\tau)d\tau^2 + r_0^2 \sinh^{-2}(r_0\tau)d\Omega^2 \right). \tag{2}$$

r_0 is the extremality parameter and satisfies [12] $r_0^2 = 2ST$, where S is the black hole entropy and T the temperature. τ is a radial coordinate and is related to the usual one r by $\sinh^{-2}(r_0\tau) = (r - r^-)(r - r^+)$. The general solution ($M$ is the black hole mass and p, q the electromagnetic charges)

$$e^{2U} = \frac{(r - r^-)(r - r^+)}{r^2}, \qquad r_0 = \sqrt{M^2 - (q^2 + p^2)} \tag{3}$$

describes the well known Reissner-Nordström non-extremal black hole and is characterized by having an outer horizon $r^+ = r_{\rm h} + r_0$ placed at $\tau \to -\infty$ and an inner one $r^- = r_{\rm h} - r_0$ at $\tau \to +\infty$. When r_0 goes to zero the two horizons end up coinciding, $\tau = -1/r$ and the solution assumes the extremal form:

$$e^{2U} = 1 - \frac{\sqrt{q^2 + p^2}}{r} = 1 - \frac{M}{r}. \tag{4}$$

An extremal charged black hole has the property to be completely defined by its charges, its temperature is vanishing but not its entropy and its near horizon geometry is a two-sphere times an anti-De-Sitter spacetime ($S^2 \otimes \mathrm{AdS}_2$).

3. The FGK formalism in Einstein-Maxwell-Scalar theories

Four-dimensional, charged, black-hole solutions for the spacetime metric can appear in theories where gravity is coupled not only to (abelian) gauge vector fields $\mathcal{A}^I$ but also to massless, neutral scalar fields ϕ^a. The Lagrangian of such a system has the general form:

$$\mathcal{L} = -R(G) + 2g_{ab}(\phi)\partial_\mu \phi^a \partial^\mu \phi^b + f_{IJ}(\phi)\mathcal{F}^I_{\mu\nu}\mathcal{F}^{J\mu\nu} + \tfrac{1}{2}h_{IJ}(\phi)\epsilon^{\mu\nu\rho\sigma}\mathcal{F}^I_{\mu\nu}\mathcal{F}^J_{\rho\sigma} \quad (5)$$

where g_{ab} is the metric of the real target space of the scalars (moduli space) and the functions $f_{IJ}(\phi)$ and $h_{IJ}(\phi)$ the couplings of the gauge fields.

By assuming spherical symmetry, using the metric ansatz (2) and the conservation of the magnetic and electric charges $\mathcal{Q} = \{p^I, q_I\}$, Ferrara, Gibbons and Kallosh in [2] showed that the correct dynamic for the remaining unfixed fields can be reproduced by the effective one-dimensional Lagrangian:

$$\mathcal{L}_{\mathrm{FGK}} = (\dot{U}(\tau))^2 + g_{ab}\dot{\phi}^a(\tau)\dot{\phi}^b(\tau) - \mathrm{e}^{2U}V_{\mathrm{bh}}(\phi, \mathcal{Q}) + r_0^2, \quad (6)$$

plus the Hamiltonian constraint:

$$(\dot{U}(\tau))^2 + g_{ab}\dot{\phi}^a(\tau)\dot{\phi}^b(\tau) + \mathrm{e}^{2U}V_{\mathrm{bh}}(\phi, \mathcal{Q}) = r_0^2. \quad (7)$$

The dot over the fields means differentiation with respect to τ. The new term V_{bh} is the so called black hole potential and it comes from the electromagnetic part of the original Lagrangian. It is a quantity quadratic in the charges and its expression depends on the scalar fields through the gauge couplings:

$$-V_{\mathrm{bh}}(\phi, \mathcal{Q}) = f^{IJ}(\phi)(q_I - h_{IK}(\phi)p^K)(q_J - h_{JK}(\phi)p^K) + f_{IJ}(\phi)p^J p^I. \quad (8)$$

The variation of (6) gives the second-order equations of motion:

$$\ddot{U} + \mathrm{e}^{2U}V_{\mathrm{bh}} = 0, \quad (9)$$

$$\ddot{\phi}^a + \Gamma^a_{bc}\dot{\phi}^b\dot{\phi}^c + g^{ab}\partial_{\phi^b}V_{\mathrm{bh}}\mathrm{e}^{2U} = 0. \quad (10)$$

By looking at them, it is evident that the black hole potential plays a key role in the evolution of the fields. In particular, in the case $r_0 = 0$, if there is for all scalars a constant ϕ_{h}^a such that

$$\partial_{\phi^a}V_{\mathrm{bh}}(\phi)\big|_{\phi_{\mathrm{h}}} = 0,$$
$$-\frac{1}{2}\partial_{\phi^a}\partial_{\phi^b}V_{\mathrm{bh}}(\phi)\big|_{\phi_{\mathrm{h}}} > 0 \qquad \forall\, a, b, \quad (11)$$

a stable extremal charged black-hole solution exists and a peculiar phenomenon, called *attractor mechanism*, takes place [13–15]. The scalars, independently from their value at spatial infinity, evolve in the moduli space as functions of the distance from the horizon till assuming on it ($\tau \to -\infty$) fixed values equal to ϕ_h^a. The solutions of (11) depend only on the charges of the black hole and for the metric one finds $r_h^2 = -V_{bh}(\phi)|_{\phi_h}$ that implies (in agreement with the no-hair theorem):

$$M^2 = -V_{bh}(\phi_h(\mathcal{Q})), \tag{12}$$

$$S = \frac{A_h}{4} = -\pi V_{bh}(\phi_h(\mathcal{Q})). \tag{13}$$

The second condition in (11) is crucial for the attractor mechanism to occur. If it is not satisfied and the extremum of (minus) the black hole potential is a maximum, the scalars must be fixed to be constant, i.e. $\phi^a(\tau) = \phi_h^a$, and the black hole is said to be unstable (each small variation of the scalars from the value extremizing V_{bh} would spoil the solution). Another possibility $(\partial_{\phi^a}\partial_{\phi^b}V_{bh}(\phi)|_{\phi_h} = 0$ for some a, b) is the existence of flat directions: the scalars on the horizon (but not the geometry of the spacetime) will in general depend on their value at infinity as well (see e.g. [16, 17]) and there will be a continuous set of critical points.

4. Static black holes in $\mathcal{N} = 2$, $D = 4$ Supergravity

The general multiplet content of ungauged $\mathcal{N} = 2$, $D = 4$ supergravity counts one supergravity multiplet, n_v vector multiplets and n_h hypermultiplets. However in the pursuit of black-hole solutions we can put to zero the fermions fields and omit the hypermultiplets. We are left in this way with the graviton, n_v complex scalars z^a and $n_v + 1$ abelian gauge fields $\mathcal{A}^I$. The Lagrangian reduces to (the bar stands for complex conjugation)

$$\begin{aligned}\mathcal{L} = &-R(G) + 2g_{a\bar{b}}(z,\bar{z})\partial_\mu z^a \partial^\mu \bar{z}^{\bar{b}} \\ &+ \Im m \mathcal{N}_{IJ}(z,\bar{z})\mathcal{F}^I_{\mu\nu}\mathcal{F}^{J\mu\nu} + \Re e \mathcal{N}_{IJ}(z,\bar{z})\epsilon^{\mu\nu\rho\sigma}\mathcal{F}^I_{\mu\nu}\mathcal{F}^J_{\rho\sigma}\end{aligned} \tag{14}$$

which is of the type of an Einstein-Maxwell-scalar theory. It follows that all the results discussed in the previous section hold true here as well and in particular, by the assumption of spherical symmetry and conservation of the charges, one can replace (14) with the effective Lagragian:

$$\mathcal{L}_{FGK} = (\dot{U}(\tau))^2 + g_{a\bar{b}}\dot{z}^a(\tau)\dot{\bar{z}}^{\bar{b}}(\tau) - e^{2U}V_{bh}(z,\bar{z},\mathcal{Q}) + r_0^2 \tag{15}$$

subject to the constraint:

$$\dot{U}^2 + g_{a\bar{b}}\dot{z}^a\dot{\bar{z}}^{\bar{b}} + e^{2U}V_{bh} = r_0^2. \tag{16}$$

What one gains in $\mathcal{N} = 2$ supergravity is that the moduli space is now a Kähler manifold and it has a richer structure with respect to the general construction above. In the cases we are interested in, it is described by the use of *special geometry* whose basis can be found for instance in [18]. The main point is that the theory is completely specified by a single holomorphic function $F = F(X^I)$. It is called prepotential and usually is expressed in terms of the homogeneous coordinates of the scalar manifold related with the affine ones by $z^a = \frac{X^a}{X^0}$. One can then define the Kähler potential $\mathcal{K} = -\ln\left[i\left(\bar{X}^I \partial_{X^I} F - X^I \overline{\partial_{X^I} F}\right)\right]$ and the covariant holomorphic symplectic section (or period vector) with Kähler weight $(1, -1)$:

$$\mathcal{V} = e^{\frac{\mathcal{K}}{2}} \begin{pmatrix} X^I \\ \partial_{X^I} F \end{pmatrix}. \tag{17}$$

The metric of the moduli space, the gauge couplings and the black hole potential are:

$$g_{a\bar{b}} = \partial_{z^a} \partial_{\bar{z}^b} \mathcal{K}, \tag{18}$$

$$\mathcal{N}_{IJ} = \overline{\partial_{X^I} \partial_{X^J} F} + 2i \frac{\Im(\partial_{X^I} \partial_{X^K} F) \Im(\partial_{X^J} \partial_{X^M} F) X^M X^K}{\Im(\partial_{X^M} \partial_{X^K} F) X^M X^K}, \tag{19}$$

$$-V_{\text{bh}} = -\frac{1}{2} Q^\Lambda Q^\Sigma \mathcal{M}_{\Lambda\Sigma}(\mathcal{N}) = -\frac{1}{2} (p^I \; q_I) \begin{pmatrix} (\mathfrak{I} + \mathfrak{R}\mathfrak{I}^{-1}\mathfrak{R})_{IJ} & -(\mathfrak{R}\mathfrak{I}^{-1})_I{}^J \\ -(\mathfrak{I}^{-1}\mathfrak{R})^I{}_J & (\mathfrak{I}^{-1})^{IJ} \end{pmatrix} \begin{pmatrix} p^J \\ q_J \end{pmatrix}, \tag{20}$$

with $\mathfrak{R}_{IJ} = \Re\mathcal{N}_{IJ}$ and $\mathfrak{I}_{IJ} = \Im\mathcal{N}_{IJ}$.

Black-hole solutions in supergravity have the additional property to preserve super-symmetry or not. Supersymmetric black holes are $\frac{1}{2}$-BPS which means [19, 20] that they conserve half of the initial supersymmetries and their mass, usually defined as $M = \dot{U}(0)$, turns out to saturate the BPS bound being equal to the absolute value at spatial infinity of the central charge

$$Z(Q, z, \bar{z}) = e^{\frac{\mathcal{K}}{2}} \left(p^I \partial_{X^I} F - q_I X^I\right). \tag{21}$$

Supersymmetry implies extremality [21] but it is not true the other way round. Non-supersymmetric black holes can indeed be extremal or non-extremal and their mass always satisfies $M > \left|Z(Q, z_\infty, \bar{z}_\infty)\right|$.

A common feature of both BPS and non-BPS black-hole solutions is the existence of first-order flow equations which imply the second-order ones obtainable from the variation of (15). The fundamental observation is that the black hole potential can be rewritten in the quadratic form

$$-e^{2U} V_{\text{bh}} = (\partial_U Y)^2 + 4g^{a\bar{b}} \partial_{z^a} Y \partial_{\bar{z}^b} Y - r_0^2 \tag{22}$$

by the introduction of a real function $Y = Y(\mathcal{Q}, z, \bar{z}, U, r_0)$ called *generalized superpotential* [22]. So, the effective Lagrangian, up to a total derivate, is equivalent to a sum of squares:

$$\mathcal{L}_{\text{FGK}} = \left(\dot{U} \pm \partial_U Y\right)^2 + \left|\dot{z}^a \pm 2g^{a\bar{b}}\partial_{\bar{z}^b}Y\right|^2 \tag{23}$$

and consequently, extremizing the respective action, leads to the first order equations (the sign is just matter of convention):

$$\dot{U} = \pm\partial_U Y, \tag{24}$$

$$\dot{z}^a = \pm 2g^{a\bar{b}}\partial_{\bar{z}^b}Y. \tag{25}$$

If $Y = e^U|Z(\mathcal{Q}, z, \bar{z})|$, the flow equations define the evolution of the fields of a supersymmetric black hole since they imply the relevant Killing spinor equations to be satisfied as well [13]. Each other possibility describes instead non-supersymmetric black holes that are extremal whenever $Y = e^U \mathcal{W}(\mathcal{Q}, z, \bar{z})$, with $\mathcal{W} \neq |Z|$ being the so called superpotential whose explicit expression depends on the model [23]. In the non-extremal case the generalized superpotential is a model-dependent complicated function which does not factorize but such that, in the limit $r_0 \to 0$, allows to obtain both the supersymmetric and non-supersymmetric extremal first-order flow equations [22, 24].

5. The H-FGK formalism

The differential equations of motion of the section above, even in the first-order formulation, are generally quite cumbersome. Till little time ago the problem of finding their solutions was approached in different ways depending on the type of black hole one was interested in. Recently, however, the perspective has changed. Given a model, the functional form of its solutions is the same for all types of black holes and expressed in terms of $2n_v + 2$ real functions $H^\Sigma(\tau)$ taken to be the new actual variables of the system [1, 25]. This is the essence of the H-FGK formalism that generalizes the idea behind the non-extremal black hole ansatz of [24, 26] and applies, possibly, to $D \geq 4$, $N \geq 2$ supergravities [27–31].

In four-dimensional $\mathcal{N} = 2$ supergravity the change of variables that takes to the H-FGK formalism is found by defining the real Kähler-neutral symplectic vectors:

$$\mathcal{R}^\Sigma \equiv \mathfrak{Re}\left(\mathcal{V}^\Sigma/\mathcal{X}\right) \equiv \tilde{H}^\Sigma, \qquad \mathcal{I}^\Sigma \equiv \mathfrak{Im}\left(\mathcal{V}^\Sigma/\mathcal{X}\right) \equiv H^\Sigma, \tag{26}$$

where $\mathcal{X} = \frac{1}{\sqrt{2}}e^{U+i\alpha}$ is a complex variable with the same Kähler weight as $\mathcal{V}^M$. The phase α does not occur in the original FGK formulation but is needed to match the degrees of freedom of the two formalisms and will lead to a gauge symmetry among the H-variables.

The capital Greek indices can be lowered and raised by the use of the symplectic metric $(\Omega_{\Sigma\Lambda}) \equiv \begin{pmatrix} 0 & 1 \\ -1 & 0 \end{pmatrix}$ and $\Omega^{\Sigma\Gamma}\Omega_{\Lambda\Gamma} = \delta^{\Sigma}{}_{\Lambda}$ according to the convention:

$$\mathcal{V}_{\Sigma} = \Omega_{\Sigma\Lambda}\mathcal{V}^{\Lambda}, \qquad \mathcal{V}^{\Sigma} = \mathcal{V}_{\Lambda}\Omega^{\Lambda\Sigma}. \tag{27}$$

The components $\mathcal{R}^{\Sigma}$ (relabeled $\tilde{H}^{\Sigma}$) can be expressed in terms of the $\mathcal{I}^{\Sigma}$ (H^{Σ}) by solving a set of algebraic equations commonly called stabilization equations [15] but to which, in light of the results of [31], the literature has begun to refer to as *Freudenthal duality equations*. The functions $\tilde{H}^{\Sigma}(H)$ are characteristic of each theory but they are always homogeneous of first degree in the H^{Σ}. By these new variables, the physical fields U and z^a, independently from the type of black hole and for all models, read:

$$e^{-2U} = W(H) \equiv \tilde{H}_{\Sigma}H^{\Sigma}, \qquad z^a = \frac{\tilde{H}^a + iH^a}{\tilde{H}^0 + iH^0}. \tag{28}$$

In the H-FGK formalism the Lagragian (15) and the Hamiltonian constraint (16) can be rewritten as [1]:

$$-I_{\text{H-FGK}}[H] = \int d\tau \left\{ \tfrac{1}{2}g_{\Sigma\Lambda}\dot{H}^{\Sigma}\dot{H}^{\Lambda} - V \right\}, \tag{29}$$

$$r_0^2 = \tfrac{1}{2}g_{\Sigma\Lambda}\dot{H}^{\Sigma}\dot{H}^{\Lambda} + V, \tag{30}$$

where we have defined the H-dependent metric

$$g_{\Sigma\Lambda} \equiv \partial_{H^{\Sigma}}\partial_{H^{\Lambda}}\log W - 2\frac{H_{\Sigma}H_{\Lambda}}{W^2} = \frac{\partial_{H^{\Sigma}}\partial_{H^{\Lambda}}W}{W} - 2\frac{H_{\Sigma}H_{\Lambda}}{W^2} - 4\frac{\tilde{H}_{\Sigma}\tilde{H}_{\Lambda}}{W^2}, \tag{31}$$

and the potential

$$V(H) = \left\{ -\tfrac{1}{4}\partial_{H^{\Sigma}}\partial_{H^{\Lambda}}\log W + \frac{H_{\Sigma}H_{\Lambda}}{W^2} \right\}\mathcal{Q}^{\Sigma}\mathcal{Q}^{\Lambda} = \left\{ -\tfrac{1}{4}g_{\Sigma\Lambda} + \tfrac{1}{2}\frac{H_{\Sigma}H_{\Lambda}}{W^2} \right\}\mathcal{Q}^{\Sigma}\mathcal{Q}^{\Lambda}. \tag{32}$$

The equations of motion are now:

$$\left[\partial_{H^{\Lambda}}\partial_{H^{\Sigma}}\log W - 2\frac{H_{\Lambda}H_{\Sigma}}{W^2} \right]\ddot{H}^{\Sigma} + \tfrac{1}{2}\partial_{H^{\Lambda}}\partial_{H^{\Sigma}}\partial_{H^{\Gamma}}\log W \left[\dot{H}^{\Sigma}\dot{H}^{\Gamma} - \tfrac{1}{2}\mathcal{Q}^{\Sigma}\mathcal{Q}^{\Gamma} \right]$$
$$-4\dot{H}_{\Lambda}\frac{\dot{H}^{\Sigma}H_{\Sigma}}{W^2} + 8H_{\Lambda}\frac{(\dot{H}^{\Gamma}\tilde{H}_{\Gamma})(\dot{H}^{\Sigma}H_{\Sigma})}{W^3} + 2\mathcal{Q}_{\Lambda}\frac{H^{\Sigma}\mathcal{Q}_{\Sigma}}{W^2} \tag{33}$$
$$-4\tilde{H}_{\Lambda}\frac{(H^{\Sigma}\dot{H}_{\Sigma})^2 + (H^{\Sigma}\mathcal{Q}_{\Sigma})^2}{W^3} = 0,$$

and the best way to treat them is to plug in an ansatz for the H^{Σ} that solves

$$\tilde{H}_{\Sigma}(\ddot{H}^{\Sigma} - r_0^2 H^{\Sigma}) + W^{-1}(\dot{H}^{\Sigma}H_{\Sigma}) = 0 \tag{34}$$

obtained by combining (30) with (33). With the additional condition

$$\dot{H}^\Sigma H_\Sigma = 0, \tag{35}$$

which ensures absence of NUT-charge, a good ansatz for the extremal case is

$$H^\Sigma = A^\Sigma - \frac{B^\Sigma}{\sqrt{2}}\tau, \tag{36}$$

while for the non-extremal extremal one

$$H^\Sigma = A^\Sigma \cosh(r_0\tau) + \frac{B^\Sigma}{r_0}\sinh(r_0\tau). \tag{37}$$

The differential equations (33) become then algebraic equations for the coefficients B^Σ while the A^Σ can be found by imposing asymptotic flatness and the definition of the scalars at infinity [24, 30]. The supersymmetric solution is always recovered when the poles of the harmonic functions (36) are $B^\Sigma = \pm Q^\Sigma$. In the non-supersymmetric case the expression of the B-coefficients is instead model-dependent and generally a function of the charges and, eventually, of the extremality parameter and the asymptotic value of the scalars.

The constraint (35) simplifies a lot the equations and it has been proved that it can be always imposed as a gauge-fixing condition of the *local Freudenthal symmetry* [31]

$$\delta_f H^\Sigma = f(\tau)\tilde{H}^\Sigma \tag{38}$$

underlying the theory for any arbitrary infinitesimal function $f(\tau)$. Rather than (35), what is actually restrictive is the harmonic/hyperbolic ansatz. Only in the extremal case it allows to explore the entire landscape of solutions while instead in the non-supersymmetric case, for certain models at least, it seems to be quite restrictive. When relaxed, different functional forms for the H^Σ may indeed be possible and in particular, for cubic models, inverse harmonic functions [32–34] are needed to correctly reproduce the extremal seed solution of [35, 36].

6. Conclusions

The problem of finding single-center, static, charged, spherically-symmetric, black-hole solutions of a generic 4-dimensional Einstein-Maxwell-scalar theory was originally simplified by the introduction of the FGK formalism. It allows to deal with just a one-dimensional effective theory whose dynamical variables are only scalar fields and, applied to $\mathcal{N} = 2$, $D = 4$ supergravity, it leaded in the past years to many advances in the study of black holes. The recent H-FGK formalism looks to promise similarly important results. It treats on the same ground all types of black-hole solutions providing manageable tools to formulate a their covariant description. The differential equations of motion, by the use of suitable ansätze, can be reduced to simpler algebraic equations and the analysis of the symmetries underlying the theory turns out to be much less involved and more intuitive. Further developments in this new formulation are expected and hopefully the problem of finding black-hole solutions in supergravity might come shortly to a complete, final understanding.

Acknowledgements

The authors are grateful to the organizers of Barcelona Postgrad Encounters on Fundamental Physics for the opportunity to contribute to the proceedings of the workshop. This work has been supported in part by grants FIS2008-06078-C03-02 and FIS2011-29813-C02-02 of Ministerio de Ciencia e Innovación (Spain) and ACOMP/2010/213 from Generalitat Valenciana.

References

[1]   P. Meessen et al., "H-FGK formalism for black-hole solutions of $\mathcal{N} = 2, d = 4$ and $d = 5$ supergravity", *Phys. Lett.* **B709** (2012), 260.

[2]   S. Ferrara, G. W. Gibbons, R. Kallosh, "Black holes and critical points in moduli space", *Nucl. Phys.* **B500** (1997), 75.

[3]   K. Schwarzschild, ""Über das Gravitationsfeld eines Massenpunktes nach der Einstein'schen Theorie"", *Sitzungsber. Kgl. Preuss. Akad. Wiss.* **1** (1916), 189.

[4]   H. Reissner, ""Über die Eigengravitation des elektrischen Feldes nach der Einstein'schen Theorie"", *Ann. Phys. (Berlin)* **50** (1916), 106.

[5]   G. Nordström, "On the Energy of the Gravitational Field in Einstein's Theory", *Verhandl. Koninkl. Ned. Akad. Wetenschap., Afdel. Natuurk., Amsterdam* **26** (1918), 1201.

[6]   E. T. Newman, A. I. Janis, "Note on the Kerr spinning particle metric", *J. Math. Phys.* **6** (1965), 915.

[7]   E. T. Newman, "Metric of a Rotating, Charged Mass", *J. Math. Phys.* **6** (1965), 918.

[8]   R. P. Kerr, "Gravitational field of a spinning mass as an example of algebraically special metrics", *Phys. Rev. Lett.* **11** (1963), 237.

[9]   J. D. Bekenstein, "Black holes and entropy", *Phys. Rev.* **D7** (1973), 2333.

[10] J. Bekenstein, "Statistical Black Hole Thermodynamics", *Phys. Rev.* **D12** (1975), 3077.

[11] J. D. Bekenstein, "Generalized second law of thermodynamics in black hole physics", *Phys. Rev.* **D9** (1974), 3292.

[12] G. W. Gibbons, R. Kallosh, B. Kol, "Moduli, scalar charges, and the first law of black hole thermodynamics", *Phys. Rev. Lett.* **77** (1996), 4992.

[13] S. Ferrara, R. Kallosh, A. Strominger, "$\mathcal{N} = 2$ extremal black holes", *Phys. Rev.* **D52** (1995), 5412.

[14] A. Strominger, "Macroscopic entropy of $\mathcal{N} = 2$ extremal black holes", *Phys. Lett.* **B383** (1996), 39.

[15] S. Ferrara, R. Kallosh, "Supersymmetry and attractors", *Phys. Rev.* **D54** (1996), 1514.

[16] S. Bellucci et al., "Charge orbits of symmetric special geometries and attractors", *Int. J. Mod. Phys.* **A21** (2006), 5043.

[17] S. Ferrara, A. Marrani, "On the Moduli Space of non-BPS Attractors for N=2 Symmetric Manifolds", *Phys. Lett.* **B652** (2007), 111.

[18] B. Craps et al., "What is special Kähler geometry?", *Nucl. Phys.* **B503** (1997), 565.

[19] B. de Wit, P. Lauwers, A. Van Proeyen, "Lagrangians of N=2 Supergravity - Matter Systems", *Nucl. Phys.* **B255** (1985), 569.

[20] A. Ceresole et al., "Duality transformations in supersymmetric Yang-Mills theories coupled to supergravity", *Nucl. Phys.* **B444** (1995), 92.

[21] J. Bellorin, P. Meessen, T. Ortín, "Supersymmetry, attractors and cosmic censorship", *Nucl. Phys.* **B762** (2007), 229.

[22] J. Perz et al., "First-order flow equations for extremal and non-extremal black holes", *JHEP* **0903** (2009), 150.

[23] A. Ceresole, G. Dall'Agata, "Flow Equations for Non-BPS Extremal Black Holes", *JHEP* **0703** (2007), 110.

[24] P. Galli et al., "Non-extremal black holes of $\mathcal{N} = 2, d = 4$ supergravity", *JHEP* **1107** (2011), 041.

[25] T. Mohaupt, O. Vaughan, "The Hesse potential, the c-map and black hole solutions", *JHEP* **1207** (2012), 163.

[26] P. Meessen, T. Ortín, "Non-Extremal Black Holes of $\mathcal{N} = 2, d = 5$ Supergravity", *Phys. Lett.* **B707** (2012), 178.

[27] P. Meessen et al., "Black holes and black strings of $\mathcal{N} = 2, d = 5$ supergravity in the H-FGK formalism", *JHEP* **1209** (2012), 001.

[28] A. de Antonio Martín, T. Ortín, C. S. Shahbazi, "The FGK formalism for black p-branes in d dimensions", *JHEP* **1205** (2012), 045.

[29] P. Bueno, R. Davies, C. S. Shahbazi, "Quantum Black Holes in Type-IIA String Theory" (2012), arXiv:1210.2817.

[30] P. Galli et al., "Black hole solutions of N=2, d=4 supergravity with a quantum correction, in the H-FGK formalism" (2012), arXiv:1212.0303.

[31] P. Galli, P. Meessen, T. Ortín, "The Freudenthal gauge symmetry of the black holes of $\mathcal{N} = 2, d = 4$ supergravity" (2012), arXiv:1211.7296.

[32] P. Galli et al., "First-order flows and stabilisation equations for non-BPS extremal black holes", *JHEP* **1106** (2011), 070.

[33] G. Bossard, S. Katmadas, "Duality covariant non-BPS first order systems", *JHEP* **1209** (2012), 100.

[34] P. Galli, K. Goldstein, J. Perz, "On anharmonic stabilisation equations for black holes" (2012), arXiv:1211.7295.

[35] G. Lopes Cardoso et al., "First-order flow equations for extremal black holes in very special geometry", *JHEP* **0710** (2007), 063.

[36] E. G. Gimon, F. Larsen, J. Simon, "Black holes in Supergravity: The Non-BPS branch", *JHEP* **0801** (2008), 040.

IV

String theory and Holography

D-branes and cosmic structure

Danielle Wills[1,a], **Konstantinos Dimopoulos**[2,b] and **Ivonne Zavala**[3,c]

[a]Centre for Particle Theory, Durham University,
South Road, DH1 3LE Durham, United Kingdom

[b]Centrum voor Theoretische Natuurkunde, Rijksuniversiteit Groningen,
Nijenborgh 4, 9747 AG Groningen, Netherlands

[c]Consortium for Fundamental Physics, Physics Department, Lancaster University,
LA1 4YB Lancaster, United Kingdom

E-mail: [1]d.e.wills@durham.ac.uk, [2]k.dimopoulos1@lancaster.ac.uk, [3]e.i.zavala@rug.nl

Abstract

We outline the embedding of the vector curvaton scenario within Type IIB string theory, where the vector field on a single D-brane plays the role of the vector curvaton, as a promising mechanism to generate statistical anisotropy. We first consider a toy model in the context of open string inflation, and then begin to construct a concrete model in the context of closed string inflation.

1. Introduction

Type IIB string theory has provided a rich arena for embedding phenomenological scenarios of the early universe within a fundamental theory, providing a wealth of light fields with masses and couplings that can in principle be known explicitly, which allows for concrete model-building. In such a context in which the early universe is expected to be diversely populated, it is natural to wonder whether or not the features captured in CMB maps are in fact the result of an intricate piece of team work, rather than a single lone wolf inflaton. In particular, the primordial seeds of cosmic structure are believed to be generated by gravitational particle production, a quantum process that requires an expanding background, but does not require the quantum and classical sectors to be one and the same. Pursuance along these lines of study led to the curvaton scenario [1] and, more recently, vector curvaton scenario [2, 3], in which the central idea is that the fields which give the dominant contribution to the seeds of cosmic structure have nothing to do with expanding the background (for vector field driven

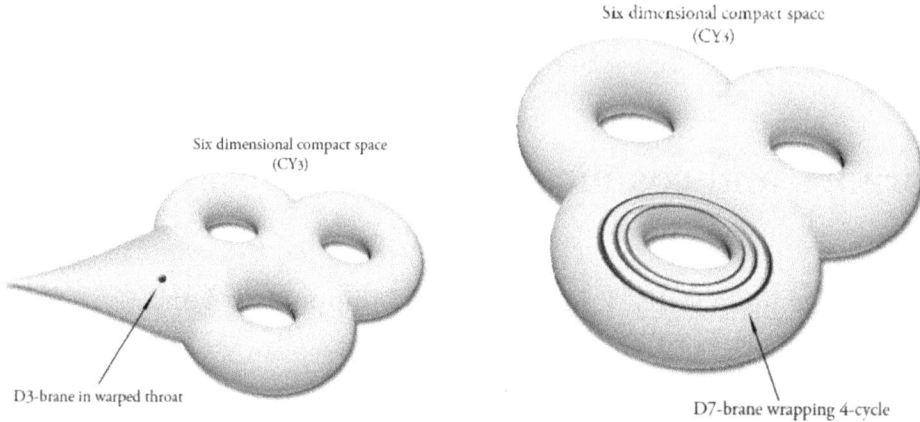

cosmologies, see e.g. [4]). An appealing attribute of the latter is that vector fields can introduce statistical anisotropy into the CMB because they naturally pick out a preferred direction.

Vector fields arise on the world-volumes of D-branes as a result of the open strings which end of them. Therefore, although their cosmological implications have been largely unexplored, such fields are completely generic in open string as well as closed string inflation models because all of these models contain D-branes, either at an active or passive level. In open string inflation, D-branes (most famously D3-branes) can actively drive cosmic expansion as they move in warped throat environments (see [5] for a review). In closed string inflation, Kähler moduli which parameterise the volumes of compact 4-cycles may drive inflation as they roll to their minima [6], and these 4-cycles may be wrapped by D7-branes, which are required to be present by consistency of the theory. In these proceedings we will explore the string constructions under which these D-brane vector fields may become curvatons.

2. The vector curvaton scenario

The seeds of cosmic structure induce tiny regions of over- and under-densities in the otherwise uniform energy density of the early universe, and at these loci the geometry is correspondingly perturbed. This gives rise to a quantity known as the curvature perturbation ζ, which forms the initial condition for structure growth and is measurable in the CMB via the Sachs-Wolfe effect. The vector curvaton scenario explores the conditions under which a vector field may produce a sizable contribution to ζ.

In order to undergo gravitational particle production, fields must be conformally non-invariant as well as sufficiently light such that their Compton wavelengths may extend beyond the horizon. In order to be viable curvatons, they must also give rise to spectra that are consistent with observations of ζ and thus, importantly, suitably scale-invariant. In one version of the vector curvaton scenario [7], it is shown that a light vector field with a mass and a gauge kinetic function that varies during inflation may give rise to a scale-invariant spectrum of superhorizon perturbations if the mass and gauge kinetic function obey

$$m \propto a(t), \qquad f \propto a(t)^2 \qquad \text{or} \tag{1}$$

$$m \propto a(t), \qquad f \propto a(t)^{-4}, \tag{2}$$

where $a(t)$ is the cosmic scale factor. To see this[1], consider the Lagrangian for a massive Abelian vector field,

$$\mathcal{L} = -\frac{1}{4} f F_{\mu\nu} F^{\mu\nu} - \frac{1}{2} m^2 A_\mu A^\mu, \tag{3}$$

where $F_{\mu\nu} = \partial_\mu A_\nu - \partial_\nu A_\mu$ is the field strength tensor, and f and m are the gauge kinetic function and mass respectively. We assume that $f \propto a(t)^c$ and $m \propto a(t)^d$, since $a(t)$ sets the only time scale in the problem. For quasi-de Sitter expansion with Hubble parameter H, we find the following equations of motion for the homogeneous $A_\mu(t)$ where $A_\mu = (A_0, \boldsymbol{A})$:

$$A_0 = 0, \tag{4}$$

$$\ddot{\boldsymbol{A}} + \left(H + \frac{\dot{f}}{f} \right) \dot{\boldsymbol{A}} + \frac{m^2}{f} \boldsymbol{A} = 0. \tag{5}$$

It is clear that the effective mass of the vector field is given by

$$M \equiv \frac{m}{\sqrt{f}}, \tag{6}$$

where we require $M \ll H$ while the cosmological scales exit the horizon. To study gravitational particle production we perturb around the homogeneous zero-mode $A_\mu(t)$,

$$A_\mu(t, \boldsymbol{x}) = A_\mu(t) + \delta A_\mu(t, \boldsymbol{x}), \tag{7}$$

and we define the canonically normalised physical (as opposed to comoving) vector field as $W = \sqrt{f} A / a$ with perturbations δW. In momentum space and for $M \ll H$, the equations of motion for the left and right transverse and longitudinal polarisations of the quantum mode functions of δW, $w(t, \boldsymbol{x})$, are found to be

$$\ddot{w}_{L,R} + 3H \dot{w}_{L,R} + \left[-\frac{1}{4}(c+4)(c-2)H^2 + \left(\frac{k}{a} \right)^2 \right] w_{L,R} = 0, \tag{8}$$

---

[1]We summarise here the computations done in [7].

$$\ddot{w}_{\parallel} + (5 - c + 2d)H\dot{w}_{\parallel} + \left[-\frac{1}{2}(c-2)(2-c+2d)H^2 + \left(\frac{k}{a}\right)^2 \right] w_{\parallel} = 0. \quad (9)$$

We may now compute the superhorizon power spectra for these polarisations. For the left and right transverse modes, we find

$$\mathcal{P}_{L,R} = \frac{k^3}{2\pi^2} \left| \lim_{k/aH \to 0^+} w_{L,R} \right|^2 \propto \left(\frac{H}{2\pi}\right)^2 \left(\frac{k}{2aH}\right)^{3-2\sigma} \quad (10)$$

where $\sigma = \frac{1}{2}|c+1|$. For the longitudinal mode,

$$\mathcal{P}_{\parallel} = \frac{k^3}{2\pi^2} \left| \lim_{k/aH \to 0^+} w_{\parallel} \right|^2 \propto \left(\frac{3H}{M}\right)^2 \left(\frac{H}{2\pi}\right)^2 \left(\frac{k}{2aH}\right)^{5-2\rho} \quad (11)$$

where $\rho = \frac{1}{2}\sqrt{9 + 2(c+1)(2-c+2d) + (2-c+2d)^2}$ and M enters via a Lorentz boost.

We then see from Eq. (10) that the transverse spectrum is scale-invariant if $c = -(1\pm 3)$. From Eq. (11), we see that the longitudinal spectrum is scale-invariant if, in addition, $d = -1/2(3 \pm 5)$, but we reject $d = -4$ because M decreases with a for this case and so $M \ll H$ cannot hold in the subhorizon limit. Hence we arrive at Eqs. (1,2). Note that a necessary condition to obtain these relations is that f and m have the precise relationships

$$f \propto m^2 \quad \text{or} \quad (12)$$
$$f \propto m^{-4}. \quad (13)$$

3. D-brane vector curvatons

The excitations of D-branes are described by the oscillation modes of the open strings that end on them. The massless bosonic open string spectrum includes a U(1) gauge boson which propagates only along the world-volume, and there are massless scalars which describe the fluctuations in the position of the brane in the transverse directions. The U(1) world-volume vector fields may then obtain masses via Stückelberg couplings to bulk two-forms, and these masses can depend on the various scalars in the theory. In addition, the vector fields may couple to the scalars through their gauge kinetic functions. If the scalars that enter these quantities are evolving at inflationary energies, then the early universe would contain massive D-brane vector fields with time varying masses and gauge kinetic functions. Then, if the vectors are light and the dynamics of the scalars obey Eqs. (1,2) while the cosmological scales exit the horizon, these D-brane vector fields could become vector curvatons and contribute to or even generate the seeds of cosmic structure. We will now examine these possibilities.

3.1. The general set-up

We consider a warped flux compactification of Type IIB string theory in which the ten dimensional metric takes the form

$$G_{MN}dx^M dx^N = h^{-1/2}g_{\mu\nu}dx^\mu dx^\nu + h^{1/2}g_{AB}dy^A dy^B, \tag{14}$$

where the indices $M,N = (0,...,9)$, $\mu,\nu = (0,...,3)$ and $A,B = (4,...,9)$ denote coordinates in ten dimensional spacetime, the four noncompact dimensions and the six compact dimensions respectively, and h is the warp factor which depends only on the compact coordinates. We embed a probe Dp-brane with world-volume coordinates ξ^a in this background, with three of its directions extended in the three large spatial dimensions, and its remaining $(p-3)$ directions wrapping a compact $(p-3)$-cycle. The action for such a brane is a sum of the Dirac-Born-Infeld (DBI) and Wess-Zumino (WZ) actions, where the former encodes kinetic terms for the brane and its world-volume fields, and the latter, couplings of world-volume fields and the brane itself to other fields in the bulk. In the Einstein frame, the DBI action takes the form

$$S_{DBI} = -\mu_p \int d^{p+1}\xi\, e^{\frac{(p-3)}{4}\phi} \sqrt{-\det(\gamma_{ab} + e^{-\frac{\phi}{2}}\mathcal{F}_{ab})}. \tag{15}$$

In this expression, ϕ is the dilaton which parameterises the string coupling, and the Dp-brane tension $T_p = \mu_p e^{((p-3)/4)\phi}$ where $\mu_p = (2\pi)^{-p}(\alpha')^{-(p+1)/2}$ and $\alpha' = \ell_s^2$ is the string scale. In addition,

$$\mathcal{F}_{ab} = \mathcal{B}_{ab} + 2\pi\alpha' F_{ab}, \qquad \gamma_{ab} = G_{MN}\partial_a x^M \partial_b x^N \tag{16}$$

where $\mathcal{B}_{ab}$ is the pullback of the NSNS 2-form field B_{ab} and F_{ab} is the field strength of the world volume gauge field, and γ_{ab} is the pullback of the ten dimensional metric on the brane.

The WZ action takes the form

$$S_{WZ} = q\mu_p \int_{\mathcal{W}_{p+1}} \sum_n \mathcal{C}_n \wedge e^{\mathcal{F}}, \tag{17}$$

where the $\mathcal{C}_n$ are the pullbacks of bulk fields C_n on the brane and $q = +1(-1)$ for a probe Dp-brane(anti-brane). The integral is over the world-volume $\mathcal{W}_{p+1}$ and the sum is over the relevant rank $(p+1)$ products of bulk and world-volume fields.

We work in the static gauge in which $\xi^\mu = x^\mu$. For the compact spacetime coordinates transverse to the brane (with indices $i,j = (p+1,...,9)$), we allow $y^i = y^i(\xi^\mu)$: these functions will give the massless open string modes that parameterise the fluctuations in the position of the brane.

3.2. D3-brane vector curvaton

We will first consider a toy model in open string inflation where the vector field on a D3-brane plays the role of the curvaton [8]. In these inflation models, the inflaton is identified with the changing position coordinate of a Dp-brane as it moves along its potential in a warped throat. If the potential is sufficiently flat, the brane moves slowly and the scalar position field which parameterises the motion has a nearly constant energy density, giving rise to an epoch of slow-roll inflation. If the potential is steep on the other hand but the brane is moving in a strongly warped region, the warping can force the brane to slow down, thus keeping the energy density roughly constant and once again giving rise to inflation, in this case known as DBI inflation [9]. The majority of models consider D3-branes however other possibilities have been studied.

Let us now focus on the vector field A_μ. From the form of Eq. (15) we see immediately that the gauge kinetic function for A_μ contains a dynamical degree of freedom, namely the dilaton ϕ. Then, for a D3-brane, the WZ action in Eq. (17) contains a coupling $C_2 \wedge F_2$ which is able to generate a mass for the vector field via the Stückelberg mechanism[2]. Computing the mass explicitly yields

$$m = e^{-\phi/2} \sqrt{\pi} (2\pi)^5 \frac{M_P}{\mathcal{V}_6}, \tag{18}$$

where $\mathcal{V}_6$ is the dimensionless volume of the compact space and M_P is the Planck mass. Therefore, both the mass and gauge kinetic function for the vector field contain the dilaton. While this field is usually considered to be stabilised during inflation, for the purpose of our toy model we will assume that it can be dynamical, at least while the cosmological scales exit the horizon.

The total four dimensional action for the D3-brane that we consider, including all appropriate terms for gravity, the evolving dilaton ϕ, the canonically normalised inflaton φ and the canonically normalised vector curvaton A_μ (after expanding the determinant in Eq. (15)[3] and taking into account the mass generation mechanism from Eq. (17)), then takes the form

$$S_{D3} = \int d^4x \sqrt{-g} h^{-1} \left(\frac{M_P^2}{2} R - \frac{M_P^2}{4} \partial_\mu \phi \partial^\mu \phi - V(\phi) \right.$$
$$\left. - \left[1 + \frac{1}{2} h e^{-\phi} F_{\mu\nu} F^{\mu\nu} + h \partial_\mu \varphi \partial^\mu \varphi \right]^{1/2} - V(\varphi) + h^{-1} - \frac{1}{2} m^2 A_\mu A^\mu \right). \tag{19}$$

---

[2]Note that for compactifications of Type IIB with O3/O7 rather than O5/O9 orientifold planes, the four dimensional components of C_2 are projected out of the spectrum, however we will continue to use this coupling for the toy model at hand.

[3]We keep terms only up to quadratic order in F and $\partial_\mu \phi$ and in their products for now, see [8] for the more general case.

We consider the fields to evolve in an FRW universe, in which case the equation of motion for $\varphi(t)$ is given by

$$\ddot{\varphi} - \frac{h'}{h^2} + \frac{3}{2}\frac{h'}{h}\dot{\varphi}^2 + 3H\dot{\varphi}\frac{1}{\gamma_\varphi^2} - \frac{h'}{h}e^{-\phi}\left(\frac{\dot{A}}{\gamma_\varphi a}\right)^2 + \left(V'(\varphi) + \frac{h'}{h^2}\right)\frac{1}{\gamma_\varphi^3} = 0, \quad (20)$$

where

$$\gamma_\varphi = \frac{1}{\sqrt{1 - h\dot{\varphi}^2}} \quad (21)$$

is the Lorentz factor for the inflaton. For a Maxwellian vector field we may expand the square root in Eq. (19) such that the equations of motion for $A_\mu(t)$ take the form given in Eqs. (4,5), with the mass m given by Eq. (18), and the gauge kinetic function f still to be determined.

We first focus on the simplest possibility: the vector field on a stationary D3-brane, in which case inflation is driven by, for example, the motion of another D3-brane. The gauge kinetic function is then

$$f = e^{-\phi}, \quad (22)$$

and the vector field decouples from the dynamics of the inflaton. Taking into account the precise powers of the dilaton ϕ that appear in Eqs. (18) and (22), we see that Eq. (12) is obtained. For our toy model we will merely assume that, in addition, Eq. (1) may hold[4], and move on to consider the cosmology, which we only briefly comment on here. The cosmological features of the vector curvaton on a stationary D3-brane have been computed in [8], where it is shown that statistical anisotropy in the power spectrum of the curvature perturbation can arise from such a scenario. The amount of statistical anisotropy, parameterised by g, is found to be

$$\sqrt{g} \sim \frac{\Omega_A}{\zeta}\frac{\delta W}{W} \sim 0.1, \quad (23)$$

where Ω_A is the density parameter for the vector field[5].

We now consider the vector field on a moving D3-brane in a warped throat. For simplicity we consider the throat to be AdS-like in the regions of interest, in which case the warp factor takes the simple form $h = \lambda/\varphi^4$ where λ is the t'Hooft coupling. In this case we find for the gauge kinetic function

$$f = e^{-\phi}\gamma_\varphi. \quad (24)$$

---

[4]This behaviour requires a linear potential for the dilaton, which could be an approximation to an exponential potential for small displacements. However, this behaviour should hold for at least the 10 e-folds that span the cosmological scales, which is unfortunately not a small displacement.

[5]Observations suggest $g \leq 0.29$ [10]. Statistical anisotropy will be observed by the Planck satellite if $g \geq 0.02$ [11].

Now, for slow-roll inflation, $\gamma_\varphi \to 1$ and the results obtained for the stationary D3-brane case follow here. For the case of DBI inflation on the other hand, it is shown in [8] that

$$f \to a(t)^{2(1+\epsilon)}, \qquad (25)$$

where ϵ is the generalised slow-roll parameter for DBI inflation. This adds a small degree of scale dependence to the spectrum, but once again the results for the stationary D3-brane follow in this case.

3.3. D7-brane vector curvaton

We will now investigate the vector field on a D7-brane in closed string inflation, and thereby construct a starting point for a concrete model of the vector curvaton scenario in string theory. Let us first outline the inflationary mechanism. We consider the effective action for $N = 1$, $d = 4$ supergravity, which takes the form

$$S = \int d^4x \sqrt{-g} \left(\frac{1}{2}R - \mathcal{K}_{i\bar{j}}\partial_\mu T^i \partial^\mu \bar{T}^{\bar{j}} - V(T^l, \bar{T}^{\bar{l}}) \right), \qquad (26)$$

where the $T_i = \tau_i + i\theta_i$ are complex Kähler chiral fields consisting of the 4-cycle volume moduli τ_i and their associated axions θ_i, and $\mathcal{K}_{i\bar{j}}$ is the Kähler metric with both holomorphic and anti-holomorphic indices, and $i,\bar{i} = 1,...,n$ where n is given by the Hodge number $h^{(1,1)}$. In the large volume scenario (LVS) [12], perturbative corrections are added to the Kähler potential and non-perturbative corrections are added to the tree level superpotential, which stabilise the Kähler moduli. The potential exhibits exponentially flat directions which support slow-roll inflation as the moduli roll to their minima [6]. For a single volume modulus τ_m for example, the potential takes the form

$$V = \frac{3W_0^2\xi}{4\mathcal{V}^3} - \frac{4W_0 a_m A_m \tau_m e^{-a_m\tau_m}}{\mathcal{V}^2}. \qquad (27)$$

Here $\mathcal{V}$ is the volume of the six dimensional Calabi-Yau (CY) in units of the string length ℓ_s, W_0 is the tree level superpotential, ξ is proportional to the Euler characteristic of the manifold, the A_m encode threshold corrections, and the a_m are constants which depend on the specific non-perturbative mechanism ($a_m = 2\pi/g_s$ for Euclidean D3-brane instantons and $a_m = 2\pi/g_s N$ for gaugino condensation on D7-branes). For a single evolving field, the canonically normalised inflaton field χ_m is related to τ_m by

$$\chi_m = \sqrt{\frac{4\lambda}{3\mathcal{V}}}\tau_m^{3/4}. \qquad (28)$$

The volume $\mathcal{V}$ and the 4-cycle moduli τ_i may be expressed in terms of the 2-cycle moduli t_i according to

$$\mathcal{V} = \frac{1}{6}\kappa_{ijk}t^i t^j t^k, \quad \tau_i = \frac{\partial}{\partial t^i}\mathcal{V} = \frac{1}{2}\kappa_{ijk}t^j t^k, \qquad (29)$$

respectively, where κ_{ijk} are the triple intersection numbers.

Geometries featuring an over-all size $\mathcal{V}$ controlled by one large 4-cycle with modulus τ_1, and then a number of small blow-up 4-cycles or "holes" with moduli $\tau_2, \ldots, \tau_n$, are referred to as "Swiss-cheese" (SC) geometries. One can often find an appropriate basis within which $\mathcal{V}$ takes a particularly simple diagonal form in terms of the 4-cycle moduli[6],

$$\mathcal{V} = \alpha \left(\tau_1^{3/2} - \sum_{i=2}^{n} \lambda_i \tau_i^{3/2} \right) \tag{30}$$

where α and λ_i are model-dependent constants. In the LVS, geometries are chosen such that $\tau_1 \gg \tau_{2,\ldots,n}$, therefore $\mathcal{V}$ is not destabilised if one (or more) of the small 4-cycles evolves.

Let us now turn to the vector field. We consider a D7-brane wrapping a 4-cycle with modulus τ, which is evolving during inflation and may be driving inflation [14]. As for the D3-brane case, the action for such a brane descends from the general DBI and WZ actions in Eqs. (15) and (17), where now we may integrate the DBI action over the internal components of the brane which wrap the compact 4-cycle. After expanding the determinant in the DBI action, this integration amounts to adding the 4-cycle modulus into the gauge kinetic function of the vector field. The WZ action for the D7-brane contains a coupling of the form

$$\frac{1}{2}(2\pi\alpha')^2 \, C_4 \wedge F_2 \wedge F_2, \tag{31}$$

where C_4 is expanded in the cohomology basis of the manifold. In this expansion is a four-dimensional 2-form D_2 which will give rise to a coupling of the form $D_2 \wedge F_2$ when one of the field strengths in Eq. (31) is a compact flux. This coupling generates a Stückelberg mass for the D7-brane vector field[7]. Again one must integrate over the compact components wrapped by the brane, and this introduces the 4-cycle modulus into the mass for the vector field. In a suitable basis, the mass is given by[8]

$$m \propto \frac{M_{\mathrm{P}}/\sqrt{\mathcal{V}}}{\sqrt{\tau^{1/2} - \frac{\tau^2}{\mathcal{V}}}} . \tag{33}$$

---

a study carried out in [13] which scanned hundreds of geometries, it is shown that a large number geometries are expressible in the "strong cheese" form in Eq. (30).

(and indeed B_2) in the D3-brane case, D_2 is not projected out of the spectrum by the action tifold planes for compactifications with O3/O7 planes.

the denominator arises from the integration over the harmonic 2-forms which yields

$$k_{ijk}t^k - \frac{(1/2 k_{ilm} t^l t^m)(1/2 k_{ino} t^n t^o)}{\mathcal{V}}, \tag{32}$$

to an appropriate basis as is used in Eq. (30).

So we see that both f and m for the D7-brane vector field contain the evolving 4-cycle modulus τ. We may now treat the dilaton as fixed and use instead the modulus τ as the varying degree of freedom relevant for the vector curvaton scenario. As such, after canonically normalising the vector field, we find

$$f = \tau. \tag{34}$$

The total four dimensional action we will consider then takes the form

$$S = \int d^4 x \sqrt{-g} \left(\frac{M_P^2}{2} R - \partial_\mu \chi \partial^\mu \chi - V(\chi) - \frac{1}{4} f F_{\mu\nu} F^{\mu\nu} - \frac{1}{2} m^2 A_\mu A^\mu \right), \tag{35}$$

where we have used Eq. (26) for a single canonically normalised inflaton χ, and the relevant terms for the D7-brane vector field from the DBI and WZ actions. Note that for closed string inflation we do not consider the D-brane scalars here.

In an FRW background, the equation of motion for $\chi(t)$ becomes

$$\ddot{\chi} + 3H\dot{\chi} + V' - \frac{1}{2} f' \left(\frac{\dot{A}}{a} \right)^2 + mm' \left(\frac{A}{a} \right)^2 = 0, \tag{36}$$

where we see that as for the D3-brane case, the vector field can backreact into the dynamics of the inflaton. The equations of motion for the vector background $A_\mu(t)$ are once again given by Eqs. (4) and (5), where now the gauge kinetic function f and mass m are given by Eqs. (34) and (33) respectively.

Having all the equations and relevant quantities at hand, we will now consider whether or not such a set-up allows for an embedding of the vector curvaton scenario. Here we will check this using simple calculations that can be done analytically. From Eq. (33) we see that for $V \gg 1$ we find $m \propto \tau^{-1/4}$, and because $f = \tau$, we immediately obtain Eq. (13). Therefore, the D7-brane curvaton can generate scale invariant spectra for all of its components if Eq. (2) holds, i.e. if $f \equiv \tau \propto a(t)^{-4}$. We will now investigate whether or not this is possible.

Let us consider the case for which the backreaction of the vector field into the dynamics of the inflaton is negligible (see [15] for a discussion of non-negligible backreaction). For $f = \tau \propto a^{-4}$, from Eq. (28) we then require that the canonically normalised field obeys $\chi \propto a(t)^{-3}$. For inflation to occur we require $V \approx const.$, and indeed the potential in Eq. (27) is exponentially flat for large enough τ. For $H \approx const.$ and using the approximation $V' \approx 0$, the solution to Eq. (36) without backreaction takes the form[9]

$$\chi = -\frac{e^{-3Ht}}{3H} c_1 + c_2,$$

---

[9]More precisely the full solution, which can only be obtained numerically using t[...] potential, will of course contain suppressed terms that come from the dependence[...] current purposes we are interested only in the most dominant contribution.

We would like $c_2 \approx 0$ in which case the dominant solution is $\chi \propto e^{-3Ht} = a^{-3}$ as required. This means that firstly the minimum of the potential should occur at a *very* small value for χ (this depends on the geometry) and secondly, χ should not approach the second term in Eq. (37) until it is very close to its minimum (this depends on initial conditions). We therefore require that initially,

$$c_1 = \dot{\chi}_0 \lesssim -3H_0\chi_0. \tag{38}$$

The value of χ should decrease during inflation, so we see that c_1 is negative as it should be. If χ is the inflaton, then we further require that $1/2\dot{\chi}^2 \ll V$ such that the energy density is roughly constant. From the Friedmann equation we obtain

$$3H_0^2\left(1 - \frac{3}{2}\dot{\chi}_0^2\right) = V(\chi), \tag{39}$$

so we require $\dot{\chi}_0 \ll \sqrt{2/3}$, i.e. the field should be sub-Planckian to begin with. Then, for 10 e-folds of evolution which span the cosmological scales, we have $N = \ln(a_f/a_i) = -1/3\ln(\chi_f/\chi_i)$, so $\chi_f \approx 10^{-14}\chi_i$, which is rather a large field range. We would further require that the other 50 e-folds of inflation take place over a very small field range, depending upon just how small the stabilised value of χ can be. This seems feasible given the vast number of Swiss cheese geometries, but it would have to be verified concretely, checking that the over-all volume and the other volume moduli are stabilised properly in the process, with realistic values for the parameters. If c_2 is a small constant, this will add a degree of dependence upon scale to the curvature perturbation, as can be seen from Eqs. (10,11) considering that $\alpha = 0$.

4. Conclusions

We have explored the possibility of embedding the vector curvaton paradigm in Type IIB string theory in both the open and closed string sectors, as a promising mechanism to generate statistical anisotropy. We first considered a toy model in open string inflation, where the vector field on a D3-brane plays the role of the vector curvaton. With simple computations we then demonstrated the feasibility of constructing a concrete model in closed string inflation, where the vector field on a D7-brane plays the role of the curvaton.

Acknowledgements

DW would like to thank Tomi Koivisto for insightful discussions. DW is supported by an STFC studentship. K.D. is supported by the Lancaster-Manchester-Sheffield Consortium for Fundamental Physics under STFC grant ST/J000418/1.

References

[1] D. H. Lyth, D. Wands, "Generating the curvature perturbation without an inflaton", *Phys. Lett.* **B524** (2002), 5.

[2] K. Dimopoulos, "Can a vector field be responsible for the curvature perturbation in the Universe?", *Phys. Rev.* **D74** (2006), 083502.

[3] K. Dimopoulos, M. Karciauskas, J. M. Wagstaff, "Vector Curvaton without Instabilities", *Phys. Lett.* **B683** (2010), 298.

[4] T. Koivisto, D. F. Mota, "Vector Field Models of Inflation and Dark Energy", *JCAP* **0808** (2008), 021.

[5] R. Kallosh, "On inflation in string theory", *Lect. Notes Phys.* **738** (2008), 119.

[6] J. P. Conlon, F. Quevedo, "Kahler moduli inflation", *JHEP* **0601** (2006), 146.

[7] K. Dimopoulos, M. Karciauskas, J. M. Wagstaff, "Vector Curvaton with varying Kinetic Function", *Phys. Rev.* **D81** (2010), 023522.

[8] K. Dimopoulos, D. Wills, I. Zavala, "Statistical Anisotropy from Vector Curvaton in D-brane Inflation", *Nucl. Phys.* **B868** (2013), 120.

[9] M. Alishahiha, E. Silverstein, D. Tong, "DBI in the sky", *Phys. Rev.* **D70** (2004), 123505.

[10] N. E. Groeneboom et al., "Bayesian analysis of an anisotropic universe model: systematics and polarization", *Astrophys. J.* **722** (2010), 452.

[11] A. R. Pullen, M. Kamionkowski, "Cosmic Microwave Background Statistics for a Direction-Dependent Primordial Power Spectrum", *Phys. Rev.* **D76** (2007), 103529.

[12] J. P. Conlon, F. Quevedo, K. Suruliz, "Large-volume flux compactifications: Moduli spectrum and D3/D7 soft supersymmetry breaking", *JHEP* **0508** (2005), 007.

[13] J. Gray et al., "Calabi-Yau Manifolds with Large Volume Vacua", *Phys. Rev.* **D86** (2012), 101901.

[14] D. Wills, "Stringy Origins of Cosmic Structure: Masterarbeit in Physik, Rheinische Friedrich-Wilhelms-Universitaet Bonn" (2011).

[15] J. M. Wagstaff, K. Dimopoulos, "Particle Production of Vector Fields: Scale Invariance is Attractive", *Phys. Rev.* **D83** (2011), 023523.

Viscosity Correlators in Improved Holographic QCD

Martin Krššák

Fakultät für Physik, Universität Bielefeld,
Postfach 100131, 33615 Bielefeld, Germany

E-mail: krssak@physik.uni-bielefeld.de

Abstract

We study a bottom-up holographic model of large-N_c Yang-Mills theory, in which conformal invariance is broken through the introduction of a dilaton potential on the gravity side. We use the model to calculate the spectral densities of the shear and bulk channels at finite temperature. In the shear channel, we compare our results to those derived in strongly coupled $\mathcal{N} = 4$ Super Yang-Mills theory as well as in weakly coupled ordinary Yang-Mills. In the bulk channel, where the conformal result is trivial, we make comparisons with both perturbative and lattice QCD. In both channels, we pay particular attention into the effects of conformal invariance breaking in the IHQCD model.

1. Introduction

Heavy ion collision data from the RHIC [1] and LHC [2] experiments suggests that quark-gluon plasma near T_c behaves as an almost ideal liquid with a very low shear viscosity to entropy ratio, $\eta/s \lesssim 0.2$. Understanding this small value has turned out to be very difficult using standard field theory methods. Calculations in perturbative QCD suggest that this ratio should be $\eta/s \approx 1$ [3], in direct contradiction with the experiments. Moreover, the shear viscosity is a real-time transport coefficient, so it is very difficult to calculate in lattice QCD.

One of the most famous predictions resulting from the AdS/CFT conjecture has been the existence of a universal bound for the viscosity to entropy ratio $\eta/s \geq 1/(4\pi)$ [4], which no physical liquid is supposed to violate. Since then, AdS/CFT methods have played an important role in increasing our understanding of strongly interacting systems, such as the matter produced in heavy ion collisions [5]. However, AdS/CFT is originally a duality between string theory and conformally invariant supersymmetric Yang-Mills theory. As QCD is not conformal nor supersymmetric, it would be important to develop a holographic model with these features, which might be able to describe real-world QCD more closely. One such model is the Improved Holographic QCD (IHQCD) proposed by Elias Kiritsis, et. al. [6, 7], where conformal invariance is broken by the introduction of a nontrivial potential for the dilaton field, mimicking perturbative QCD in the UV region and giving the model a linear glueball spectrum in the IR, i.e. making it confining.

In this talk, we present results published in the papers [8, 9], where using IHQCD we have calculated the field theory correlators

$$G_s^R(\omega, k = 0) = -i \int d^4x \, e^{i\omega t} \theta(t) \langle [T_{12}(t, x), T_{12}(0, 0)] \rangle \tag{1}$$

$$G_b^R(\omega, k = 0) = -i \int d^4x \, e^{i\omega t} \theta(t) \left\langle \left[\frac{1}{3} T_{ii}(t, x), \frac{1}{3} T_{jj}(0, 0) \right] \right\rangle, \tag{2}$$

where T_{12} and T_{ii} are the shear and bulk components of the energy-momentum tensor, respectively. Of our primary interest are the spectral densities, which are defined as

$$\rho_{s,b}(\omega, T) = \Im \, G_{s,b}^R(\omega, k = 0). \tag{3}$$

Using the Kubo formulas, we can furthermore formulate the relations

$$\eta = \lim_{\omega \to 0} \frac{\rho_s(\omega, T)}{\omega}, \tag{4}$$

$$\zeta = \lim_{\omega \to 0} \frac{\rho_b(\omega, T)}{\omega} \tag{5}$$

between the spectral functions and the corresponding transport coefficients, the shear (η) and bulk (ζ) viscosities of the theory.

As IHQCD is a two-derivative model, we expect the shear viscosity to entropy ratio to have the universal value $\eta/s = 1/(4\pi)$, as proven in general in [10]. This is in fact one of our reasons to focus on the properties of the spectral densities, as they allow us to distinguish between strongly coupled $\mathcal{N} = 4$ SYM and the field theory dual to IHQCD. In the shear channel, we furthermore have available a recent pertubative result for the Yang-Mills spectral density [11], which naturally is not applicable in the region of low frequencies, but provides us with an important consistency check at higher frequencies. In the bulk channel, the SYM spectral density (and bulk viscosity) on the other hand vanish due to conformal invariance, so a comparison of our IHQCD results with the perturbative result of [12] and the lattice data of [13] (for Euclidean imaginary time correlators) is of utmost interest.

2. Improved Holographic QCD

Improved Holographic QCD is a five-dimensional bottom-up holographic model, where on the gravitational side one starts with a gravity+dilaton system and introduces a potential for the dilaton field to describe some crucial features of conformal symmetry breaking [6, 7]. The action reads

$$S = \frac{1}{16\pi G_5} \int \mathrm{d}^5 x \sqrt{-g} \left[R - \frac{4}{3} \left(\frac{\partial \lambda}{\lambda} \right)^2 + V(\lambda) \right],\tag{6}$$

where we have used $\lambda = e^\phi$, with ϕ denoting a dilaton field. The form of the potential $V(\lambda)$ is found by matching the holographic beta function to the 2-loop perturbative one (in large-N_c Yang-Mills theory) and by requiring the model to possess a linear glueball spectrum. This can be achieved, for example, through the potential [14]

$$V(\lambda) = \frac{12}{\mathcal{L}^2} \left[1 + \frac{88}{27}\lambda + \frac{4619}{729}\lambda^2 \frac{\sqrt{1 + \ln(1 + \lambda)}}{(1 + \lambda)^{2/3}} \right].\tag{7}$$

Note that this is the potential we have used in our most recent publication [9], while in [8] a slightly different choice, consistent with the original potential of [6, 7], was made. For possible issues in the derivation of the potential we would like to point on [15, 16].

A background metric consistent with the above potential reads

$$\mathrm{d}s^2 = b^2(z) \left[-f(z)\,\mathrm{d}t^2 + \mathrm{d}x^2 + \frac{\mathrm{d}z^2}{f(z)} \right],\tag{8}$$

where the functions $f(z)$, $b(z)$ and $\lambda(z)$ are determined from the Einstein equations (6)

$$\dot{W} = 4bW^2 - \frac{1}{f}(W\dot{f} + \frac{1}{3}bV),\tag{9}$$

$$\dot{b} = -b^2 W,\tag{10}$$

$$\dot{\lambda} = \frac{3}{2}\lambda\sqrt{b\dot{W}},\tag{11}$$

$$\ddot{f} = 3\dot{f}\,bW.\tag{12}$$

An additional constraint is that the space-time be asymptotically AdS, i.e. that close to the boundary

$$b(z) \to \frac{\mathcal{L}}{z}, \qquad z \to 0.\tag{13}$$

where $\mathcal{L}$ is the curvature radius of the AdS space-time, which without loss of generality we can set to $\mathcal{L} = 1$. Using eqs. (9)–(12), we can next determine the thermodynamic properties of the system [17, 18]. Matching these results to the corresponding quantities in large-N_c Yang-Mills theory, we are able to finally fix the value of the gravitational constant

$$\frac{\mathcal{L}^3}{4\pi G_5} = \frac{4N_c^2}{45\pi^2}\,.$$

(14)

To determine correlators of the energy-momentum tensor on the field theory side, we follow [19, 20] and introduce perturbations around the background metric (8),

$$g_{00} = b^2 f\left(1+\epsilon H_{00}\right),\ g_{11} = b^2\left(1+\epsilon H_{11}\right),\ g_{55} = \frac{b^2}{f}\left(1+\epsilon H_{55}\right),\ g_{12} = \epsilon b^2 H_{12}\,.$$

(15)

Here, ϵ is a power counting parameter, and the fluctuation H_{12} corresponds to the T_{12} operator, while the fluctuation H_{11} is dual to $\frac{1}{3}T_{ii}$. Expanding next the Einstein equations to the first order in ϵ, we obtain fluctuation equations for the perturbations

$$\ddot{H}_{12} + \frac{\mathrm{d}}{\mathrm{d}z}\log(b^3 f)\dot{H}_{12} + \frac{\omega^2}{f^2}H_{12} = 0\,,$$

(16)

$$\ddot{H}_{11} + \frac{\mathrm{d}}{\mathrm{d}z}\log(b^3 f X^2)\dot{H}_{11} + \left(\frac{\omega^2}{f^2} - \frac{\dot{f}\dot{X}}{fX}\right)H_{11} = 0\,,$$

(17)

where

$$X(\lambda) \equiv \frac{\dot{\lambda}}{3\lambda\dot{b}/b}\,.$$

(18)

These equations are to be solved using infalling boundary conditions at the horizon [9].

Having obtained the solutions to the above fluctuation equations, the spectral functions are available using standard results [19],

$$\rho_s(\omega, T) = \frac{s(T)}{4\pi}\frac{\omega}{\left|H_{12}(z\to 0)\right|^2}\,,$$

(19)

$$\rho_b(\omega, T) = 6X_h^2\frac{s(T)}{4\pi}\frac{\omega}{\left|H_{11}(z\to 0)\right|^2}\,.$$

(20)

In the latter bulk channel result, X_h denotes the quantity of eq. (18) evaluated at the horizon. When comparing the holographic results to those of lattice QCD, another useful observable turns out to be the directly measurable Euclidean imaginary time correlator. This quantity is obtained from the spectral density through the relation

$$G(\tau, T) = \int_0^\infty \frac{\mathrm{d}\omega}{\pi}\rho(\omega, T)\frac{\cosh\left[\left(\frac{\beta}{2}-\tau\right)\pi\omega\right]}{\sinh\left(\frac{\beta}{2}\omega\right)}\,,\quad \beta \equiv 1/T\,.$$

(21)

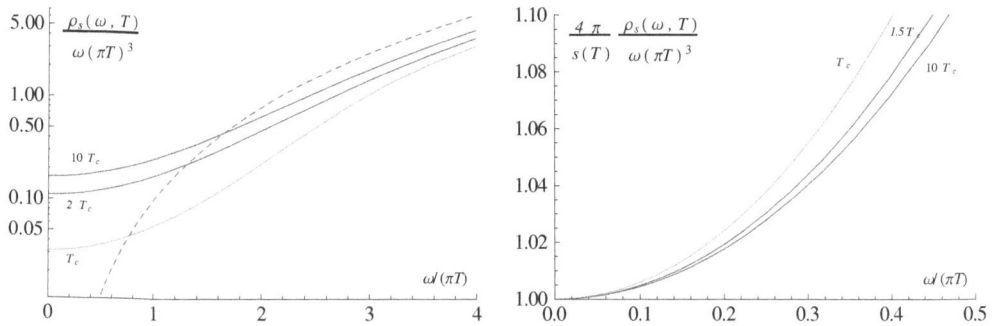

Figure 1.: Left: the ratio of the IHQCD spectral density $\rho_s(\omega)$ and frequency ω in the shear channel, displayed for three different temperatures in units of $\mathcal{L}^3/(4\pi G_5)$. The dashed curve corresponds to the asymptotic limit of $\pi\omega^3/32$. Right: the ratio of the spectral density and frequency normalized by the entropy. Using the Kubo formula of eq. (4), we observe that the shear viscosity over entropy obtains the universal value of $1/(4\pi)$.

3. Results

Below, we present our results for the spectral densities, covering the cases of the shear and bulk channels separately.

3.1. Shear Channel

The ase of the shear channel was considered already in [8], which, unlike [9], however did not make use of the potential of eq. (7). Thus, we will here concentrate on the results for the latter reference, showing in fig. 1 (left) the ratio of the spectral density and ω for various temperatures. We observe a behavior similar to that displayed in fig. 3 of [8]; however, due to the choice of potential the logarithmic convergence of our result towards its large-ω limit ($\rho_{as} = \pi\omega^4/32$ in units where $\mathcal{L}^3/(4\pi G_5)$ has been set to unity) is somewhat slower. Analogous calculations in $\mathcal{N} = 4$ SYM theory (cf. [21]) show that for temperatures close to the critical one, a significant difference between the IHQCD and SYM results arises, which however rapidly disappears at higher temperatures.

1 (right), we next plot ρ_s/ω normalized by the entropy, from where we can read e of the shear viscosity over entropy ratio can as the intercept of the curves As a consequence of IHQCD being a two-derivative model, we find the ediction of $\eta/s = 1/(4\pi)$ to hold at all temperatures considered [10].

mitations of perturbative QCD in the region of low frequencies, at high l behavior of spectral densities is expected to reduce to its predictions

195

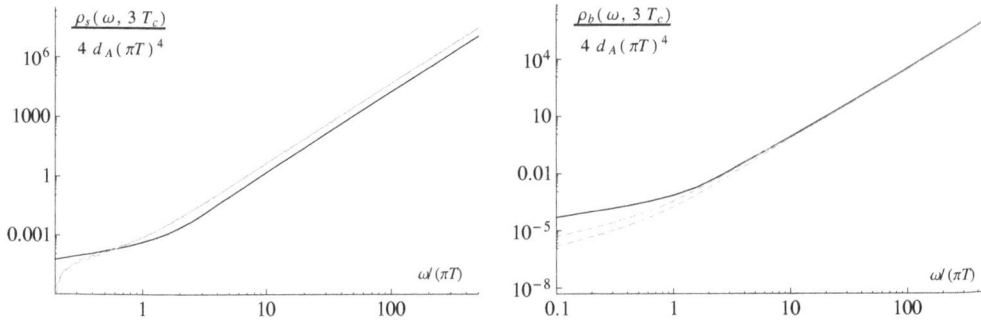

Figure 2.: Spectral densities in the shear (left) and bulk (right) channels for $T = 3T_c$, normalized by $4d_A = 4(N_c^2 - 1)$. The black curves correspond to the IHQCD and the dashed dashed curves to the 2-loop perturbative QCD results [11, 12].

due to asymptotic freedom [11, 22, 23]. To this end, in fig. 2 (left), we plot both the perturbative (dashed red curve) and IHQCD (black curve) results for the shear spectral density, finding that in the UV limit, the IHQCD spectral function has the same functional behaviour as the perturbative one up to some overall normalization constant. As explained in detail in [9], this proportionality factor can be identified as 4/9 by analytically determining the asymptotic limits of both the perturbative and holographic spectral densities.

3.2. Bulk Channel

Next, we move on to present the results of our recent correlator calculations in the bulk channel [9]. In fig. 3 (left), we first plot $\rho_b(\omega)/\omega$ for different temperatures, normalized by the factor $4d_A$. Using the Kubo formula (5), the bulk viscosity can again be read off as the value of the curves in the limit $\omega \to 0$. From these results, we see that the bulk viscosity decreases with increasing temperature, ultimately approaching the vanishing conformal limit. In fig. 2 (right), we next find that in the region of large ω, the IHQCD spectral function reduces to the perturbative QCD prediction. This reduction to is even more apparent when we study ratio the ρ_b/ω^4 plotted in fig. 3 (right), where we see that this ratio quickly approaches the temperature independent limit $\propto 1/(\log \omega/T_c)^2$. This form in fact coincides with the ω-dependence of the field theory gauge coupling g^4, to which the leading order perturbative result is proportional to [12, 24].

Finally, lattice simulations of course play an important role in the study interacting phenomena, as they are the only fundamentally nonperturbati ciples method available. However, due to technical reasons only Euclidea are possible to measure on the lattice, and to this end we next speciali of imaginary time correlators, cf. eq. (21). With this result, it is possibl

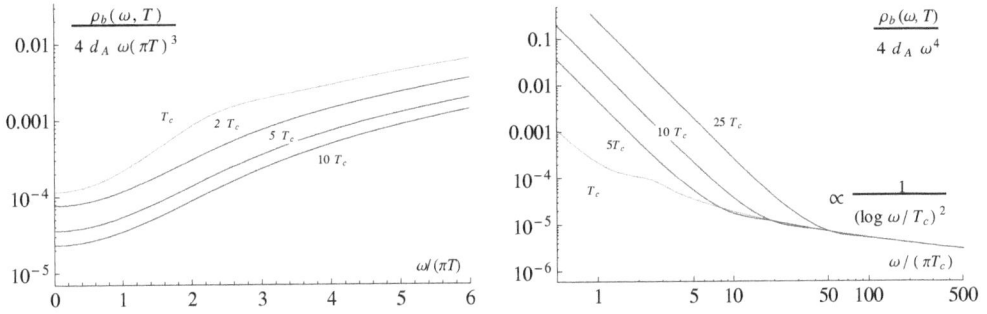

Figure 3.: Left: the ratio of the IHQCD bulk spectral density $\rho_b(\omega)$ and ω, normalized by $4d_A$. Right: the ratio ρ_b/ω^4, plotted as a function of $\omega/(\pi T_c)$ for multiple temperatures, normalized by $4d_A$. For large values of ω, this ratio reduces to the temperature independent limit $1/(\log \omega/T_c)^2$.

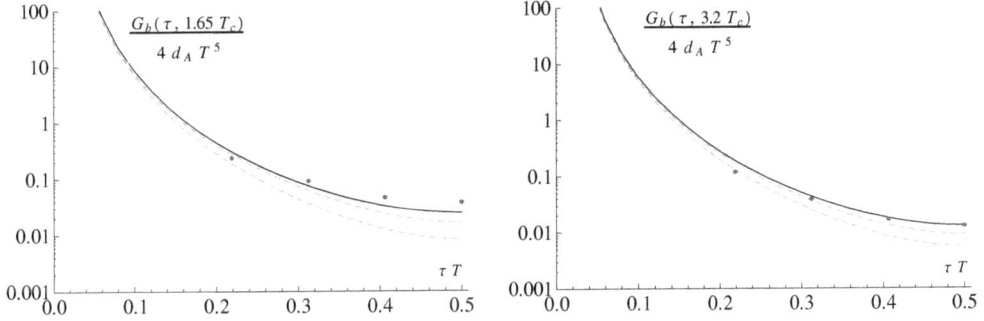

Figure 4.: The imaginary time correlators of the bulk channel, computed for two different temperatures in IHQCD (black curves) and pQCD (red dashed curves) [12] and compared with the lattice data (blue points) of [13].

Euclidean correlators in the both IHQCD and pQCD, and furthermore compare the results with the lattice data of [13], see fig. 4. We find that the holographic results seem to be in better agreement with the lattice data than the perturbative ones for all temperatures considered.

4. Conclusions

In this talk, we have presented the results of a recent calculation of finite temperature correlation functions in both the shear and bulk channels of strongly coupled large-N_c Yang-Mills theory. After motivating our research and stressing the importance of using non-conformal holographic models to describe the physics of QCD, we briefly introduced the fundamental properties of the IHQCD model and showed how spectral densities can be obtained in this framework.

Next, we compared our holographic results with corresponding calculations in strongly coupled $\mathcal{N} = 4$ supersymmetric Yang-Mills theory [21], perturbative Yang-Mills theory [11, 12] as well as lattice simulations in the same theory [13]. In the shear channel, we observed a significant difference between the IHQCD and SYM results close to the critical temperature T_c, which however rapidly disappears at higher temperatures. We also compared the IHQCD and perturbative spectral densities in the large-ω region, finding in both cases an asymptotic ω^4 behavior, albeit with differing overall normalizations.

Similar methods were next used to calculate the IHQCD spectral density in the bulk channel, where we confirmed that with increasing temperature, the bulk viscosity approaches the vanishing result known to occur in conformal theories. Subsequently, we compared our holographic calculation with results from perturbative Yang-Mills theory, finding impressive agreement in the UV (large-ω) limit, including a perfectly matching overall normalization. We find this observation somewhat puzzling, considering the qualitatively different conclusion reached in the shear channel.

Finally, in the bulk channel, we calculated imaginary time correlators using both the IHQCD and perturbative QCD spectral functions, and compared the results with available lattice data. We conclude that the lattice data seems to favour our holographic results over perturbative one for all temperatures considered.

Acknowledgements

The author would like to thank Keijo Kajantie and Aleksi Vuorinen for collaboration, and in addition Yan Zhu for sharing her perturbative QCD results in a useful form as well as Harvey Meyer for providing his lattice data. The work was supported by the DFG graduate school *Quantum Fields and Strongly Interacting Matter* as well as the Sofja Kovalevskaja programme of the Alexander von Hulboldt foundation. I would also like to express my gratitude to the organizers of this conference for hospitality and for the possibility to present this talk.

References

[1]   M. Tannenbaum, "Highlights from BNL-RHIC" (2012), arXiv:1201.5900.

[2]   B. Muller, J. Schukraft, B. Wyslouch, "First Results from Pb+Pb collisions at the LHC", *Ann. Rev. Nucl. Part. Sci.* **62** (2012), 361.

[3]   P. B. Arnold, G. D. Moore, L. G. Yaffe, "Transport coefficients in high temperature gauge theories. 2. Beyond leading log", *JHEP* **0305** (2003), 051.

[4]   P. Kovtun, D. Son, A. Starinets, "Viscosity in strongly interacting quantum field theories from black hole physics", *Phys. Rev. Lett.* **94** (2005), 111601.

[5] J. Casalderrey-Solana et al., "Gauge/String Duality, Hot QCD and Heavy Ion Collisions" (2011), arXiv:1101.0618.

[6] U. Gursoy, E. Kiritsis, "Exploring improved holographic theories for QCD: Part I", *JHEP* **0802** (2008), 032.

[7] U. Gursoy, E. Kiritsis, F. Nitti, "Exploring improved holographic theories for QCD: Part II", *JHEP* **0802** (2008), 019.

[8] K. Kajantie et al., "Frequency and wave number dependence of the shear correlator in strongly coupled hot Yang-Mills theory", *Phys. Rev.* **D84** (2011), 086004.

[9] K. Kajantie, M. Krssak, A. Vuorinen, "Energy momentum tensor correlators in hot Yang-Mills theory: holography confronts lattice and perturbation theory" (2013), arXiv:1302.1432.

[10] N. Iqbal, H. Liu, "Universality of the hydrodynamic limit in AdS/CFT and the membrane paradigm", *Phys. Rev.* **D79** (2009), 025023.

[11] Y. Zhu, A. Vuorinen, "The shear channel spectral function in hot Yang-Mills theory" (2012), arXiv:1212.3818.

[12] M. Laine, A. Vuorinen, Y. Zhu, "Next-to-leading order thermal spectral functions in the perturbative domain", *JHEP* **1109** (2011), 084.

[13] H. B. Meyer, "The Bulk Channel in Thermal Gauge Theories", *JHEP* **1004** (2010), 099.

[14] M. Jarvinen, E. Kiritsis, "Holographic Models for QCD in the Veneziano Limit", *JHEP* **1203** (2012), 002.

[15] K. Veschgini, E. Megias, H. Pirner, "Trouble Finding the Optimal AdS/QCD", *Phys. Lett.* **B696** (2011), 495.

[16] E. Megias, H. Pirner, K. Veschgini, "Thermodynamics of AdS/QCD within the 5D dilaton-gravity model", *Nucl. Phys. Proc. Suppl.* **207-208** (2010), 333.

[17] U. Gursoy et al., "Holography and Thermodynamics of 5D Dilaton-gravity", *JHEP* **0905** (2009), 033.

[18] J. Alanen et al., "Mass spectrum and thermodynamics of quasi-conformal gauge theories from gauge/gravity duality", *Phys. Rev.* **D84** (2011), 086007.

[19] S. S. Gubser, S. S. Pufu, F. D. Rocha, "Bulk viscosity of strongly coupled plasmas with holographic duals", *JHEP* **0808** (2008), 085.

[20] U. Gursoy et al., "Thermal Transport and Drag Force in Improved Holographic QCD", *JHEP* **0912** (2009), 056.

[21] K. Kajantie, M. Vepsalainen, "Spatial scalar correlator in strongly coupled hot N=4 Yang-Mills theory", *Phys. Rev.* **D83** (2011), 066003.

[22] S. Caron-Huot, "Asymptotics of thermal spectral functions", *Phys. Rev.* **D79** (2009), 125009.

[23] Y. Schroder et al., "The Ultraviolet limit and sum rule for the shear correlator in hot Yang-Mills theory", *JHEP* **1112** (2011), 035.

[24] M. Laine, M. Vepsalainen, A. Vuorinen, "Ultraviolet asymptotics of scalar and pseudoscalar correlators in hot Yang-Mills theory", *JHEP* **1010** (2010), 010.

Holographic Cutoff Flow of Anomalous transport coefficients

Amadeo J. Alba[1] and Luis Melgar[2]

Instituto de Física Teórica, Universidad Autónoma de Madrid,
C/ Nicolás Cabrera 13-15, Cantoblanco, 28049 Madrid, Spain

E-mail: [1]amadeo.j@gmail.com, [2]luis.melgar@csic.es

Abstract

We study anomalous induced transport at finite temperature and chemical potential at strong coupling. To do so we use a holographic bottom-up model that induces the desired global anomalies in the dual four dimensional CFT. Due to the topological nature of the anomalies involved, we expect the anomalous transport coefficients not to receive corrections beyond one loop, as well as to present a trivial cutoff flow. We check explicitly these two points using the holographic model.

1. Introduction and overview

Anomalies appear in the context of relativistic quantum field theories. In four dimensions, chiral anomalies [1] involve triangle diagrams with either vector currents only or vector currents and the energy momentum tensor, in which case one speaks of a (mixed gauge-)gravitational anomaly [2]. They are responsible for the breakdown of a classical symmetry due to quantum effects. If the symmetry is local, anomalies impose severe restrictions on the structure and definition of gauge theories (for comprehensive reviews on anomalies see [3]). In the case of a symmetry generated by a set of generators T_A, and considering only right-handed fermions, the presence of a chiral anomaly in vacuum is encoded in a non-vanishing $d_{ABC} = \frac{1}{2} \text{Tr} \left(T_A \{T_B, T_C\} \right)$. The corresponding parameter in the case of the gravitational anomaly is $b_A = \text{Tr} \left(T_A \right)$. Recently, it has been pointed out that, at finite temperature and density, anomalies are responsible for the appearance of new non-dissipative transport phenomena [4–6]: The Chiral Magnetic Effect (CME) and the Chiral Vortical Effect (CVE). The history of these effects goes back to the early eighties, when their manifestations were already shown in the physics of neutrinos [7], although it was not until 2008 when the name "Chiral Magnetic Effect" was written for the first time [8].

In the CME, an external magnetic field induces a current parallel to it

$$J^\mu = \sigma_{\rm B} B^\mu, \qquad\qquad \sigma_B = \frac{\mu_{\rm A}}{4\pi^2}, \qquad\qquad (1)$$

where $\sigma_{\rm B}$ is the chiral magnetic conductivity and $B^\mu = \frac{1}{2}\epsilon^{\mu\nu\rho\lambda}u_\nu F_{\rho\lambda}$. The axial chemical potential $\mu_{\rm A}$ may be generated by a mechanism that has been proposed in [8]. A straightforward way to compute the conductivity coefficient is to calculate the electric current of chiral fermions in the presence of a constant magnetic field. By doing this, one finds that the zeroth Landau level is chiral and that it is the only one that contributes to the current and therefore an imbalance in chirality (signaled by the non-zero axial chemical potential) leads to the appearance of an electric current. The role of the anomaly comes precisely from the coefficient of this conductivity, which has been shown to be proportional to d_{ABC}.

The CVE refers to the creation of a current parallel to the vortices in the fluid

$$J^\mu = \sigma_{\rm V}\omega^\mu, \qquad\qquad \sigma_{\rm V} = \frac{\mu_{\rm A}^2 + \mu^2}{8\pi^2} + \frac{T^2}{12}, \qquad\qquad (2)$$

with $\omega^\mu = \epsilon^{\mu\nu\rho\lambda}u_\nu\partial_\rho u_\lambda$ being the vorticity vector and u_μ the fluid four-velocity. Its most surprising property comes from the temperature dependent term in the conductivity (2), since its coefficient has been shown to coincide with the mixed axial-gravitational anomaly polynomial in four dimensions [9]. This relationship, however, remains obscure in some sense, since the physical microscopic mechanism which relates both effects is not properly understood yet (see however [10]).

Remarkably enough, it was found that the form of the anomalous conductivities can be derived purely from Ward Identities of a theory with an Energy Momentum tensor and a current[1] [12]. Moreover, anomalous transport coefficients are dissipationless, since they belong to the anti-hermitian part of the retarded Green's functions [9]. This implies that their effect can be seen even at equilibrium. Within the framework of Hydrodynamics, the actual calculation of the coefficients was initially carried out by imposing the divergence of the entropy current to be positive definite [13]. This has the disadvantage that the microscopic derivation of the coefficients remains unclear. Later on, several approaches have been used to calculate the anomalous transport without making use of this local version of the second [12, 14, 15].

The contribution of the gravitational anomaly to these transport coefficients was first obtained in a weakly (actually, zero) coupled gas of chiral fermions in [16] and some subtleties regarding the definition of a chemical potential for an anomalous symmetry

---

[1]Up to an arbitrary coefficient [11] that is related to the Mixed anomaly [9].

were addressed in several articles [17, 18]. In a beautiful work [19], R. Laganayagam and P. Surówka generalized [16] by studying anomalous transport in an ideal Weyl gas. By writing down a conservation equation for the chiral spectral current and solving it, the authors found a generalization of the anomalous conductivities valid for any (even) dimension and an expression that relates the anomaly induced transport coefficients to the anomaly polynomial of the Ideal Weyl gas. Furthermore, in [20], a definition for the local entropy current for higher-curvature gravitational theories was proposed and the Fluid/Gravity correspondence was applied to compute the first order conductivities in the presence of the gravitational anomaly. The expression for the second order transport coefficients (including the anomalous ones) is now under study [21]. Recently, in [22] an observable called thermal helicity has been defined. It has been postulated that, at finite temperature and density and under certain assumptions, this observable can be derived entirely from the anomaly polynomial and therefore could serve to study generalizations of the analysis made so far.

We analyze transport properties of relativistic fluids driven by anomalies. Due to the topological nature of anomalies, we expect them not to depend on the details of the interaction. Here we try to understand this fact by focusing on the strong coupling limit of the theory[2]. Thus, for our purposes, the gauge/gravity duality represents a very useful tool. We will be using a "bottom-up" model which implements the anomalies holographically by means of the Kubo formulae[3] [24]

$$\sigma_{\mathrm{B}} = \lim_{p_n \to 0} \frac{i}{2p_c} \sum_{a,b} \epsilon_{abc} \langle J^a J^b \rangle (\omega = 0, \boldsymbol{p}), \tag{3}$$

$$\sigma_{\mathrm{V}} = \lim_{p_n \to 0} \frac{i}{2p_c} \sum_{a,b} \epsilon_{abc} \langle J^a T_0^b \rangle (\omega = 0, \boldsymbol{p}). \tag{4}$$

The related transport coefficients for the energy current $T^{0\mu} = \sigma_{\mathrm{B}}^{\epsilon} B^{\mu}$, $T^{0\mu} = \sigma_{\mathrm{V}}^{\epsilon} \omega^{\mu}$, can be calculated via

$$\sigma_{\mathrm{B}}^{\epsilon} = \lim_{p_n \to 0} \frac{i}{2p_c} \sum_{a,b} \epsilon_{abc} \langle T_0^a J^b \rangle (\omega = 0, \boldsymbol{p}), \tag{5}$$

$$\sigma_{\mathrm{V}}^{\epsilon} = \lim_{p_n \to 0} \frac{i}{2p_c} \sum_{a,b} \epsilon_{abc} \langle T_0^a T_0^b \rangle (\omega = 0, \boldsymbol{p}). \tag{6}$$

In what follows we will compute the anomalous conductivities and check whether they coincide with the results at weak coupling [25]. In the second part of this work we study the cutoff flow of the anomalous transport coefficients, showing that, as expected and after several subtleties that are addressed, the flow is trivial.

---

[2]We refer the interested reader to the references given above to find a weak coupling approach.

[3]Latin letters denote purely spatial indices here. A computation using the Fluid/Gravity correspondence, instead of the Kubo formulae, can be found in [23].

1.1. Holographic computation of anomalous conductivities

In this section we make an overview of how the CME and CVE are implemented holographically. The first step towards a holographic computation of anomalous conductivities is to set the appropriate anomalies in the dual theory. Since it has been found that the CVE receives contributions not only from the axial anomaly, but from the mixed axial-gravitational anomaly as well, a term that induces this anomaly at the boundary must be added too:

$$S_{ACS} = \frac{\lambda}{16\pi G} \int_{r<\Lambda} d^5x \sqrt{-g}\, \epsilon^{MNPQR} A_M R^A{}_{BNP} R^B{}_{AQR} . \tag{7}$$

The complete action reads

$$S = \frac{1}{16\pi G} \int_{r<\lambda} d^5x \sqrt{-g} \left[R + 2\Lambda_c - \frac{1}{4} F_{MN} F^{MN} \right] + S_{ACS} + S_{AEM} + S_\partial + S_{CSK}, \tag{8}$$

where

$$S_{AEM} = \frac{\kappa}{48\pi G} \int d^5x \sqrt{-g}\, \epsilon^{MNPQR} A_M F_{NP} F_{QR} , \tag{9}$$

$$S_\partial = -\frac{1}{8\pi G} \int_\partial d^4x \sqrt{-h} K , \tag{10}$$

$$S_{CSK} = -\frac{\lambda}{2\pi G} \int_{\partial\mathcal{M}} d^4x \sqrt{-h} n_M \epsilon^{MNPQR} A_N K_{PL} D_Q K^L_R . \tag{11}$$

S_{AEM} is the term that implements the axial anomaly. S_∂ is the usual Gibbons-Hawking term. S_{CSK} is added for physical reasons, since it ensures that the anomalous Ward identity for gauge transformations depends only on the intrinsic curvature tensor at $r = \Lambda$ [9]. In this section, the attention is focused on the limit $\Lambda \to \infty$; later on we will study the dependence of the conductivities with the cutoff.

Applying the well-known AdS/CFT dictionary [26], one can compute the one-point function of the current and check the anomalous Ward identity explicitly. Indeed, the *covariant*[4] current turns out to be

$$16\pi G J^A = n_B \left[F^{AB} - 8\epsilon^{BACDE} \lambda K_{CF} D_D K^E_F \right]_{r=\Lambda} \tag{12}$$

with a purely four dimensional divergence that on shell evaluates to

$$D_\mu J^\mu = -\frac{1}{16\pi G} \epsilon^{opqr} \left[\frac{\kappa}{3} F_{op} F_{qr} + \lambda R^a{}_{(4)bop} R^b{}_{(4)aqr} \right]_{r=\Lambda} \tag{13}$$

---

[4]The interested reader is referred to [24] for a discussion regarding the consistent/covariant definition of the anomalous current in holography.

where $\epsilon^{opqr} \equiv \epsilon^{nopqr}$ is the four dimensional epsilon tensor. The usual expression for the mixed anomaly [2] is thus recovered if we include S_{CSK}. To evaluate the on-shell action requires solving the equations of motion (EOM) for all fields:

$$G_{MN} - \Lambda_c g_{MN} = \frac{1}{2} F_{ML} F_N{}^L - \frac{1}{8} F^2 g_{MN} + 2\lambda \epsilon_{LPQR(M} \nabla_B \left(F^{PL} R^B{}_{N)}{}^{QR} \right), \qquad (14)$$

$$\nabla_N F^{NM} = -\epsilon^{MNPQR} \left(\kappa F_{NP} F_{QR} + \lambda R^A{}_{BNP} R^B{}_{AQR} \right). \qquad (15)$$

The background solution we will be working with is an asymptotically anti-de Sitter black hole that fixes generic (homogeneous) one-point functions:

$$ds^2 = \frac{r^2}{L^2} \left(-f(r) dt^2 + dx^2 \right) + \frac{L^2}{r^2 f(r)} dr^2, \qquad A^{(0)} = \left(\beta - \frac{\mu r_H^2}{r^2} \right) dt. \qquad (16)$$

The horizon of the black hole is located at $r = r_H$ and the blackening factor is $f(r) = 1 - \frac{ML^2}{r^4} + \frac{Q^2 L^2}{r^6}$. The dual state corresponds to a thermal equilibrium ensemble at finite temperature and chemical potential. The chemical potential of the theory is *defined* here as the energy necessary to bring a unit charge from the horizon to infinity: $\mu \equiv A(r \to \infty) - A(r = r_H)$. Henceforth, the arbitrary constant β is set to zero. The parameters M, Q are related to the chemical potential at infinity, μ, and r_H by $M = \frac{r_H^4}{L^2} + \frac{Q^2}{r_H^2}$, $Q = \frac{\mu r_H^2}{\sqrt{3}}$. Finally, the Hawking temperature is given by $T = \frac{2r_H^2 M - 3Q^2}{2\pi r_H^5}$.

Remarkably enough, the asymptotically AdS background (16) is also a solution to the Einstein-Maxwell equations (i.e. when $\kappa = \lambda = 0$), which in particular means that it does not excite the Chern-Simons terms (the anomalies). The effect of having $\kappa \neq 0$, $\lambda \neq 0$ is hence only seen at first order in perturbations of the gauge and gravitational fields. We can therefore say that the Chern-Simons terms in the action correspond here to relevant deformations of the UV theory and are treated perturbatively.

Now we proceed to compute the conductivities of the CME and CVE from the Kubo formulae (3)–(6). The 2-point correlators of the currents and the energy momentum tensor can be obtained holographically from the action to second order in perturbations. Without loss of generality we consider the perturbations on top of the background solution with momentum k in the y-direction and at zero frequency. It is only necessary to turn on the shear sector, that is, the perturbations are written as $a_\alpha(r) e^{iky}$, $h_t^\alpha(r) e^{iky}$, where $\alpha = x, z$.[5] It is more convenient to work with the coordinate $u = \frac{r_H^2}{r^2}$ instead of r (in this sense, we define $u_c \equiv r_H^2/\Lambda^2$). The equations of motion for the perturbations in

---

[5] At zero frequency the fields h_y^α decouple from the system and thus will not be considered (see [9]).

the shear sector, with $\omega = 0$ and to $\mathcal{O}(k)$, read

$$0 = h_t^{\alpha''}(u) - \frac{h_t^{\alpha'}(u)}{u} - 3auB_\alpha'(u) + i\bar\lambda k \epsilon_{\alpha\beta} \left[\left(24au^3 - 6(1 - f(u))\right) \frac{B_\beta(u)}{u} \right.$$

$$\left. +(9au^3 - 6(1 - f(u)))B_\beta'(u) + 2u(uh_t^{\beta'}(u))' \right], \tag{17}$$

$$0 = B_\alpha''(u) + \frac{f'(u)}{f(u)}B_\alpha'(u) - \frac{h_t^{\alpha'}(u)}{f(u)}$$

$$+ ik\epsilon_{\alpha\beta} \left(\frac{3}{uf(u)} \bar\lambda \left(\frac{2}{a}(f(u) - 1) + 3u^3\right) h_t^{\beta'}(u) + \bar\kappa \frac{B_\beta(u)}{f(u)} \right), \tag{18}$$

where $\bar\lambda = \frac{4\mu\lambda L}{r_{\rm H}^2}$, $\bar\kappa = \frac{4\mu\kappa L^3}{r_{\rm H}^2}$, $B_\alpha = a_\alpha/\mu$ and $a = \mu^2 L^2/(3r_{\rm H}^2)$. The above equations display a crucial property: even though (14)–(15) contain three and four derivatives, (17)–(18) happen to be second order in u-derivatives. This implies that a solution is found by specifying only two boundary conditions (BC). According to the holographic dictionary, the computation of retarded green functions requires imposing infalling rather than outgoing conditions at the horizon, the latter corresponding to the calculation of advanced green functions. So we will impose infalling conditions at the horizon[6] and fix the boundary values of the perturbations (Dirichlet BC).

The second order on-shell action can be written in a rather compact form, in terms of the transform of the vector of perturbations $\Phi_k^\top(u) = \left(B_x(u), h_t^x(u), B_z(u), h_t^z(u)\right)$:

$$\delta S^{(2)} = \int\limits_{r=\Lambda} \frac{d^d k}{(2\pi)^d} \left[\Phi_{-k}^I \mathcal{A}_{IJ} \Phi_k^{'J} + \Phi_{-k}^I \mathcal{B}_{IJ} \Phi_k^J \right] \tag{19}$$

where the matrices $\mathcal{A}$ and $\mathcal{B}$ depend only on the background. The expression for the Green functions can be written as $G_{IJ}(k, u_c) = -2 \lim_{u \to u_c} \left(\mathcal{A}_{IM} \left(F_J^M(k, u)\right)' + \mathcal{B}_{IJ} \right)$, where $F_J^I(k, u)$, normalized so that $F_J^I(k, u = 0) = \mathbb{1}$, is the *bulk-to-boundary propagator* $\Phi_k^I(u) = F_J^I(k, u)\phi_k^J$ and $\phi_k^J \equiv \Phi_k^I(u = 0)$ corresponds to the *source* of the dual operator. In order to get the explicit quantities, we need to solve the EOM of the perturbations. Expanding the perturbations to first order in momentum, we find an analytically solvable system (for a more detailed explanation, see [9]). Once we have the solution to the equations, the correlators can be evaluated and the conductivities subsequently determined by means of the Kubo formulae (3)–(6):

$$\sigma_{\rm B} = \frac{\mu}{4\pi^2}, \qquad \sigma_{\rm V} = \sigma_{\rm B}^\epsilon = \frac{\mu^2}{8\pi^2} + \frac{T^2}{24}, \qquad \sigma_{\rm V}^\epsilon = \frac{\mu^3}{12\pi^2} + \frac{\mu T^2}{12}. \tag{20}$$

---

[6]Regularity of the gauge fields as well as vanishing metric fluctuations at the horizon.

2. Flow of the anomalous transport coefficients

One would expect that the remarkable fact that both the CME and the CVE are completely determined by anomalies, which are very robust (topological) objects, prevents these transport coefficients from acquiring corrections beyond one-loop, as strongly sugested by [9]. For the same reasons, it would be in principle expectable that no dependence on the RG group scale or the cutoff scale is found. However, recent studies indicate that there are indeed corrections to the CVE arising from couplings to dynamical gauge fields [13, 27].

Within the gauge-gravity duality, the running with the holographic coordinate can be interpreted as a type of renormalization group (RG) flow in the dual field theory [28]. The first application of this holographic flow to transport coefficients is [29], where it was shown that the electric conductivity and the shear viscosity have a trivial flow. The extension to finite chemical potential has been studied in [30], and in [31] the flow is analyzed in the framework of the Gauß-Bonnet theory.

Recently, there is a renewed interest in this subject due to the explicit holographic construction of the Wilsonian Renormalization Group [32]. This approach has also served to study in detail the holographic dual of the cutoff scale [33].

Let us study whether the anomalous transport coefficients suffer from corrections due to the cutoff scale at very strong coupling. In order to do so, we will use holographic techniques and methods which resemble the ones used firstly by Iqbal and Liu [29] to compute the flow of non-anomalous transport coefficients. The model is exactly the same as the one used in the previous section, but with a finite cutoff Λ implemented. Several technical remarks are in order here:

- The model corresponds to a definition of a theory equipped with a cutoff Λ, being the bulk action (7) directly defined at $r < \Lambda$; this means that we are not considering a (holographic) Wilsonian partition function [32], but rather one in which the UV degrees of freedom have been neglected. The interpretation of (7) as a generating functional of a cutoff theory is present already in the first works on holographic RG [28]. The resulting theory is non-local at the scale of the cutoff.

- No counterterm (for a general explanation of the necessity of counterterms in holography, see [34]) has been included because the anomalous transport coefficients are finite. Furthermore, the contribution of the counterterms is at least of order k^2, and therefore negligible if we restrict ourselves to first order in k. In particular, this means that the renormalized conjugate momenta will not be recovered after removing the cutoff.

- Studying correlators as functions of Λ corresponds to analyzing the value of the Green's functions for *different* theories, each one equipped with a cutoff.

This is why the forthcoming analysis does not correspond to a proper RG flow, but to a cutoff flow. Several equivalent methods to compute the flow of the anomalous transport coefficients are detailed in [35]. Here we will focus on the most powerful one, namely, the one in which the flow of the anomalous transport coefficients is seen as the flow of the retarded conductivities.

More concretely, the flow can be determined by considering the system living between the horizon and a cutoff surface placed at $r = \Lambda$. The 2-point functions computed at the hypersurface (which plays the role of the "boundary"), making use of the holographic dictionary, are interpreted as the 2-point functions in a theory equipped with a cutoff Λ. It is not straightforward to establish what type of regularization procedure in the Quantum Field Theory side is dual to this setup, for such a map has not been constructed yet. For instance, the naive guess of a hard cutoff regularization is discarded because Lorentz invariance is preserved in the holographic hypersurface. Therefore, it is not in principle easy to perform this flow analysis in the QFT side.

In order to compute the correlators, we follow essentially the prescription of [26], as before. However, we now have to normalize the bulk-to-boundary propagator so that $F_J^I(k, u_c) = \mathbb{1}$ and the boundary condition at u_c is $\Phi_k^I(u_c) = \phi_k^I$, which is the (coarse-grained) source defined at Λ.

There is, however, a non-trivial issue that must be addressed at this point. As aforementioned, the Mixed anomaly introduces several complications due to the fact that the λ-term (7) contains derivatives of order higher than two. This in turn implies that the equations of motion (14)–(15) are of higher order in derivatives, so we need more than two boundary conditions to completely fix a solution. Hence, it is not surprising that the variation of the higher order terms $S_{ACS} + S_{CSK}$ leads to a boundary term involving the variation of the extrinsic curvature $\delta K_{ij} : -\frac{\lambda}{2\pi G} \int_\partial \sqrt{-h} \epsilon^{mlqr} D_r A_m \delta K_q^\nu K_{l\nu}$. Such a term does not allow us to directly apply the holographic dictionary and obtain a *general* expresion for the energy momentum tensor as the variation of the on-shell action with respect to the (induced) metric (i.e. roughly $\langle T^{ij} \rangle \sim \delta S^{\text{on-shell}} / \delta \gamma_{ij}$). In order to determine the spectrum of operators and get expressions for generic n-point functions, it would be necessary to consider the full higher derivative theory, in which, in particular, K_{ij} should be in principle considered to be a new coordinate independent from the canonical momentum conjugated to the metric.

Here we will adopt a much more practical point of view: even though we are forbidden to write down generic expressions for n-point functions, we can manage to compute two-point functions involving the current and the energy-momentum tensor for the vacuum (16). Notice that, to second order in perturbations, the boundary contribution reads

$$-\frac{\lambda}{2\pi G} \int_\partial \sqrt{-h} \epsilon^{mlqr} D_r \delta A_m \delta K_q^\nu K_{l\nu} , \tag{21}$$

and other possible terms would vanish when evaluated in the background solution. The fact that the equations for the perturbations are second order in derivatives becomes very important here, because we can completely determine the solution and then evaluate the boundary contribution (21) on it. For the shear sector and in the limit we are interested in, $\delta K_q^v \propto \dot{h}_q^v$ does not need to be provided as a boundary condition and is completely determined by the solution h_q^v, once we specify Dirichlet conditions at u_c and infalling BC at the horizon.

3. Final results and Conclusion

Now we are in position to apply the prescription of [26]. The retarded two-point function $G_{IJ}(k, u_c)$ is found after taking the variation of the on-shell action twice with respect to the sources. The contribution of (21) is taken into account in the definition of the $\mathcal{A}, \mathcal{B}$ matrices (for details, see [35]). Observe that the limit is not taken to infinity, but to the cutoff. The resulting 2-point functions are

$$\langle J^x J^z \rangle = -\mathrm{i}k \frac{\mu(u_c)}{4\pi^2}, \qquad \langle J^x T_t^z \rangle = \mathrm{i}k \left(\frac{\mu^2(1 - u_c)^2}{8\pi^2} + \frac{T^2}{24} \right), \qquad (22)$$

$$\langle T_t^x J^z \rangle = \mathrm{i}k \left(\frac{\mu^2(1 - u_c)^2}{8\pi^2} + \frac{T^2}{24} \right), \quad \langle T_t^x T_t^z \rangle = -\mathrm{i}k \left(\frac{\mu^3(1 - u_c)^3}{12\pi^2} + \frac{\mu(\Lambda)T^2}{12} \right). \tag{23}$$

Recall that, once we know the two-point functions, we can apply the Kubo formulae (3)–(6) and compute the anomalous conductivities. Moreover, it is straightforward to verify that equations (22)–(23) are compatible with the asymptotic value (20), computed in [9], after we take $u_c \to 0$ (or $\Lambda \to \infty$).

There is a physical consistency check of the prescription used to treat the problematic boundary term. Since a_α and $h_{\alpha\beta}$ are the sources for J^α and $T^{\alpha\beta}$ respectively, (21) only contributes to the correlators $\langle T_t^x J^z \rangle$, $\langle J^x T_t^z \rangle$. We can thus apply the well-known program to compute $\langle T_t^x T_t^z \rangle$ and $\langle J^x J^z \rangle$, without worrying about (21), and then impose that, whatever method we use to deal with (21), the results we get for $\langle T_t^x J^z \rangle$ and $\langle J^x T_t^z \rangle$ are compatible with the one obtained for the two-point functions involving only two currents or two energy-momentum tensors. Here, by *compatibility* we mean that, for physical reasons, the cutoff flow of the chemical potential and the temperature present in all the correlators must be the same. Taking a look at (22)–(23) we check that $\langle T_t^x T_t^z \rangle$ and $\langle J^x J^z \rangle$ suggest the existence of a flowing chemical potential $\mu(\Lambda) \equiv \mu(1 - u_c)$ and a non-flowing (Hawking) temperature; and $\langle T_t^x J^z \rangle$, $\langle J^x T_t^z \rangle$, computed using the prescription detailed above, confirm this Λ-dependence in a non-trivial way.

So, even though a Dirichlet boundary condition is no longer enough to define the variational problem properly, we have used a heuristic prescription that allows to compute 2-point functions (without discussing general definitions of the corresponding operators). This procedure, which can be seen to be the most natural one by using physical arguments, yields 2-point functions that are consistent and whose flows do not get in contradiction with the result found in the absence of gravitational anomaly. In view of the topological nature and the non-renormalization theorem for the chiral magnetic conductivity, it is at first sight somewhat surprising to find a non-trivial flow (although several unexpected corrections have been found recently [13], [27]). This flow becomes however natural if we define the chemical potential in its elementary way as the energy needed to introduce one unit of charge into the ensemble. In the holographic dual this corresponds to bringing a unit of charge from the boundary, now situated at $r = \Lambda$ behind the horizon. The energy difference between a unit of charge at the boundary and a unit of charge at the horizon is just given by $A_0(\Lambda) - A_0(r_H) = \left(\beta - \frac{\mu r_H^2}{\Lambda^2} \right) - (\beta - \mu) = \mu(\Lambda)$. This defines an effective chemical potential in the theory equipped with the cutoff Λ and is precisely the quantity that appears in our flow equations. In fact, the definition of such an effective chemical potential is natural even in field theory. If we have a momentum cutoff of order Λ, we can localize a unit charge only inside a volume within a radius of order $1/\Lambda$. Thermalizing this unit of charge means spreading it out over the entire ensemble. The difference in energy between the two configurations, the unit of charge localized within $1/\Lambda$ and the one spread out over the ensemble is again the effective chemical potential. On the other hand, the temperature does not flow. The temperature of the hole ensemble is given by the Hawking temperature T and it depends only on the radius of the horizon r_H, the radius of AdS and the chemical potential at infinity. In conclusion, The CME can be expressed in the form

$$\sigma(\Lambda) = \frac{N_c \mu(\Lambda)}{2\pi^2} . \tag{24}$$

It is linear in the chemical potential and the numerical coefficient is independent of the cutoff. In this sense, it obeys the expected non-renormalization theorem. Identical arguments can be applied to the rest of the anomalous conductivities.

Acknowledgements

The authors would like to thank the organizers of the Workshop "Barcelona Postgrad Encounters on Fundamental Physics" and K. Landsteiner for fruitful discussions. This work has been supported by Plan Nacional de Altas Energías FPA2009-07908 and FPA2008-01430, CPAN (CSD2007-00042), Comunidad de Madrid HEP-HACOS S2009/ESP-1473. L.M. has been supported by fellowship BES-2010-041571. A.J.A is supported through the FPU grant AP2010-5686.

References

[1]   S. L. Adler, "Axial vector vertex in spinor electrodynamics", *Phys. Rev.* **177** (1969), 2426.

[2]   R. Delbourgo, A. Salam, "The gravitational correction to pcac", *Phys. Lett.* **B40** (1972), 381.

[3]   R. Bertlmann, "Anomalies in Quantum Field Theory", Oxford, UK: Clarendon 566 p. (International series of monographs on physics: 91) ()1996.

[4]   M. A. Metlitski, A. R. Zhitnitsky, "Anomalous axion interactions and topological currents in dense matter", *Phys. Rev.* **D72** (2005), 045011.

[5]   G. Newman, D. Son, "Response of strongly-interacting matter to magnetic field: Some exact results", *Phys. Rev.* **D73** (2006), 045006.

[6]   D. Son, A. R. Zhitnitsky, "Quantum anomalies in dense matter", *Phys. Rev.* **D70** (2004), 074018.

[7]   A. Vilenkin, "Macroscopic parity violating effects: neutrino fluxes from rotating black holes and in rotating thermal radiation", *Phys. Rev.* **D20** (1979), 1807.

[8]   K. Fukushima, D. E. Kharzeev, H. J. Warringa, "The Chiral Magnetic Effect", *Phys. Rev.* **D78** (2008), 074033.

[9]   K. Landsteiner et al., "Holographic Gravitational Anomaly and Chiral Vortical Effect", *JHEP* **1109** (2011), 121.

[10]   K. Jensen, R. Loganayagam, A. Yarom, "Thermodynamics, gravitational anomalies and cones" (2012), arXiv:1207.5824.

[11]   Y. Neiman, Y. Oz, "Relativistic Hydrodynamics with General Anomalous Charges", *JHEP* **1103** (2011), 023.

[12]   K. Jensen, "Triangle Anomalies, Thermodynamics, and Hydrodynamics", *Phys. Rev.* **D85** (2012), 125017.

[13]   D. T. Son, P. Surowka, "Hydrodynamics with Triangle Anomalies", *Phys. Rev. Lett.* **103** (2009), 191601.

[14]   N. Banerjee et al., "Constraints on Fluid Dynamics from Equilibrium Partition Functions", *JHEP* **1209** (2012), 046.

[15]   K. Jensen et al., "Towards hydrodynamics without an entropy current", *Phys. Rev. Lett.* **109** (2012), 101601.

[16]   K. Landsteiner, E. Megias, F. Pena-Benitez, "Gravitational Anomaly and Transport", *Phys. Rev. Lett.* **107** (2011), 021601.

[17]   K. Landsteiner, E. Megias, F. Pena-Benitez, "Anomalies and Transport Coefficients: The Chiral Gravito-Magnetic Effect" (2011), arXiv:1110.3615.

[18]   K. Landsteiner et al., "Gravitational Anomaly and Hydrodynamics", *J. Phys. Conf. Ser.* **343** (2012), 012073.

[19] R. Loganayagam, P. Surowka, "Anomaly/Transport in an Ideal Weyl gas", *JHEP* **1204** (2012), 097.

[20] S. Chapman, Y. Neiman, Y. Oz, "Fluid/Gravity Correspondence, Local Wald Entropy Current and Gravitational Anomaly", *JHEP* **1207** (2012), 128.

[21] E. Megias, F. Pena-Benitez, "Second order transport coefficients with anomalies" (*In preparation*).

[22] R. Loganayagam, "Anomalies and the Helicity of the Thermal State" (2012), arXiv:1211.3850.

[23] S. Bhattacharyya et al., "Nonlinear Fluid Dynamics from Gravity", *JHEP* **0802** (2008), 045.

[24] I. Amado, K. Landsteiner, F. Pena-Benitez, "Anomalous transport coefficients from Kubo formulas in Holography", *JHEP* **1105** (2011), 081.

[25] D. E. Kharzeev, H. J. Warringa, "Chiral Magnetic conductivity", *Phys. Rev.* **D80** (2009), 034028.

[26] M. Kaminski et al., "Holographic Operator Mixing and Quasinormal Modes on the Brane", *JHEP* **1002** (2010), 021.

[27] F. Hou, H. Liu, H.-c. Ren, "A Possible Higher Order Correction to the Vortical Conductivity in a Gauge Field Plasma", *Phys. Rev.* **D86** (2012), 121703.

[28] V. Balasubramanian, P. Kraus, "Space-time and the holographic renormalization group", *Phys. Rev. Lett.* **83** (1999), 3605.

[29] N. Iqbal, H. Liu, "Universality of the hydrodynamic limit in AdS/CFT and the membrane paradigm", *Phys. Rev.* **D79** (2009), 025023.

[30] Y. Matsuo, S.-J. Sin, Y. Zhou, "Mixed RG Flows and Hydrodynamics at Finite Holographic Screen", *JHEP* **1201** (2012), 130.

[31] X.-H. Ge et al., "Holographic RG Flows and Transport Coefficients in Einstein-Gauss-Bonnet-Maxwell Theory", *JHEP* **1201** (2012), 117.

[32] T. Faulkner, H. Liu, M. Rangamani, "Integrating out geometry: Holographic Wilsonian RG and the membrane paradigm", *JHEP* **1108** (2011), 051.

[33] S. Grozdanov, "Wilsonian Renormalisation and the Exact Cut-Off Scale from Holographic Duality", *JHEP* **1206** (2012), 079.

[34] K. Skenderis, "Lecture notes on holographic renormalization", *Class. Quant. Grav.* **19** (2002), 5849.

[35] K. Landsteiner, L. Melgar, "Holographic Flow of Anomalous Transport Coefficients", *JHEP* **1210** (2012), 131.